Frontiers in Mathematics

More information about this series at http://www.springer.com/series/5388

Rainer Picard • Des McGhee • Sascha Trostorff •
Marcus Waurick

A Primer for a Secret
Shortcut to PDEs
of Mathematical Physics

 Birkhäuser

Rainer Picard
Institut für Analysis
TU Dresden
Dresden, Germany

Des McGhee
Department of Mathematics and Statistics
University of Strathclyde
Glasgow, UK

Sascha Trostorff
Mathematisches Seminar
Christian-Albrechts-Universität zu Kiel
Kiel, Germany

Marcus Waurick
Department of Mathematics and Statistics
University of Strathclyde
Glasgow, UK

ISSN 1660-8046   ISSN 1660-8054   (electronic)
Frontiers in Mathematics
ISBN 978-3-030-47332-7   ISBN 978-3-030-47333-4   (eBook)
https://doi.org/10.1007/978-3-030-47333-4

Mathematics Subject Classification: 35F, 35-01

This book is published under the imprint Birkhäuser, www.birkhauser-science.com, by the registered company Springer Nature Switzerland AG.
The registered company address is: Gewerbestrasse 11, 6330 Cham, Switzerland

Introduction

A typical entry point into the field of (linear) partial differential equations is to consider general polynomials $P(\partial)$ in $\partial := (\partial_0, \ldots, \partial_n)$ with (complex or real) matrix coefficients. Here ∂_k denotes the partial derivative with respect to the variable in the position labelled with[1] $k \in \{0, \ldots, n\}$, $n \in \mathbb{N}$. Even if we discuss solutions only in the whole Euclidean space $\mathbb{R}^{n+1}$ the solution theory for an equation of the form

$$P(\partial) u = f$$

involving a general partial differential operator $P(\partial)$ is quite involved and one quickly restricts attention to very specific polynomials. Indeed, the equations relevant to applications are not that varied. One commonly investigates three subclasses, loosely labelled as elliptic, parabolic, and hyperbolic, to present specific solution methods for each of them.

However, when viewed from the right perspective there is a single subclass containing these three types (and many more), which can be characterized conveniently and solved with one and the same method. To explain the corresponding rigorous framework is the objective of this text.

The theory we will present in this book is rooted in [57], with some first generalizations to be found in [55,59]. We shall refer also to [63,79,83,87,88] for generalizations towards nonlinear or non-autonomous setups. The interested reader will find a more detailed survey in [68, 75]. In the present book, however, we shall present the core yet surprisingly elementary solution theory for what we will call *evolutionary equations*.

The structure of this class of partial differential expressions can be formally described by two matrices[2] $M_0, M_1 \in \mathbb{R}^{(N+1)\times(N+1)}$, $N \in \mathbb{N}$. The partial differential operator

---

[1] Note that we usually prefer to start our numbering with 0. In particular, $\mathbb{N}$ denotes the set of non-negative integers.

[2] Indeed, keeping in mind that a complex number $x + iy$ can be understood as a (2×2)-matrix of the form

$$\begin{pmatrix} x & -y \\ y & x \end{pmatrix},$$

where $x, y \in \mathbb{R}$, we may actually assume that M_0 and M_1 have only real entries.

$P(\partial)$ will then be assumed to be of the form

$$P(\partial) = \partial_0 M_0 + M_1 + A\left(\widehat{\partial}\right), \tag{1}$$

where $A\left(\widehat{\partial}\right)$ denotes a polynomial in $\widehat{\partial} := (\partial_1, \ldots, \partial_n)$, that is, borrowing jargon from applied fields, only in the "spatial" variables, if we consider ∂_0 to be the derivative with respect to "time". In this terminology, if we focus on "relevant" partial differential equations, we may focus on first-order-in-time systems. Moreover, in standard cases we have structural features of $P(\partial)$ which narrow down the class of differential operators even further. We assume[3]

$$A^*\left(-\widehat{\partial}\right) = -A\left(\widehat{\partial}\right)$$

and

$$M_0 = M_0^* \text{ and } \mathfrak{Re}\, M_1 := \frac{1}{2}\left(M_1 + M_1^*\right) \geq c_0 > 0. \tag{2}$$

In applications, the latter positive definiteness constraint is rarely satisfied. However, after a simple formal transformation[4] we get

$$\partial_0 M_0 + \widetilde{M}_1 + A\left(\widehat{\partial}\right) = \exp\left(-\varrho m_0\right)\left(\partial_0 M_0 + M_1 + A\left(\widehat{\partial}\right)\right)\exp\left(\varrho m_0\right)$$

---

[3] With this, $A\left(\widehat{\partial}\right)$ becomes skew-selfadjoint in $L^2\left(\mathbb{R}^n\right)$ and—by canonical extension to the time-dependent case—in $L^2\left(\mathbb{R}^{1+n}\right)$. If $A\left(\widehat{\partial}\right) = \sum_{\alpha \in \mathbb{N}^n} A_\alpha \widehat{\partial}^\alpha$, then $A\left(\widehat{\partial}\right)^* = A^*\left(-\widehat{\partial}\right) := \sum_{\alpha \in \mathbb{N}^n} A_\alpha^* \left(-\widehat{\partial}\right)^\alpha$ and this constraint means that the matrix coefficients A_α, $\alpha = (\alpha_1, \ldots, \alpha_n)$, are selfadjoint or skew-selfadjoint depending on the order $|\alpha| := \sum_{k=1}^n \alpha_k$ being even or odd, respectively. Note that since $A\left(\widehat{\partial}\right)$ is a polynomial, only finitely many of the coefficients are non-vanishing. In most cases, the maximal order is actually also just 1.

[4] This transformation shifts the rigorous functional analytical discussion from $L^2\left(\mathbb{R}^{n+1}\right)$ to the more appropriate setting in the Hilbert space $H_{\varrho,0}\left(\mathbb{R}; L^2\left(\mathbb{R}^n\right)\right)$, which is defined such that

$$\exp\left(-\varrho m_0\right) : H_{\varrho,0}\left(\mathbb{R}; L^2\left(\mathbb{R}^n\right)\right) \to L^2\left(\mathbb{R}; L^2\left(\mathbb{R}^n\right)\right) = L^2\left(\mathbb{R}^{1+n}\right)$$

$$\varphi \mapsto \exp\left(-\varrho m_0\right)\varphi$$

becomes a unitary mapping. Here the multiplication operator $\exp\left(-\varrho m_0\right)$ is defined via $\left(\exp\left(-\varrho m_0\right)\varphi\right)(t) := \exp\left(-\varrho t\right)\varphi(t)$, $t \in \mathbb{R}$. We will be more precise and detailed later.

where

$$\widetilde{M}_1 := M_1 - \varrho M_0.$$

Now, the constraints (2) translate to

$$M_0 = M_0^* \text{ and } \varrho M_0 + \mathfrak{Re}\, M_1 \geq c_0 > 0 \qquad (3)$$

and the latter strict positive definiteness constraint needs to hold only for all sufficiently large $\varrho \in\,]0, \infty[$. As we shall see later, the particular role of time is encoded in this bias for *positive* values of parameter ϱ.

To improve on the range of applicability, we will generalize the above problem class by allowing M_0 and M_1 to be Hilbert space operators and A to be a general skew-selfadjoint operator so that operators of the "space-time" form

$$\partial_0 M_0 + M_1 + A \qquad (4)$$

can be treated. In the proper setting, ∂_0 will be seen to be a continuously invertible operator, which, among other things, allows us to consider the operator $M\left(\partial_0^{-1}\right) :=$ $M_0 + \partial_0^{-1} M_1$, which in application occurs when describing so-called material laws. We therefore shall refer to $M\left(\partial_0^{-1}\right)$ as well as to M_0 and M_1 as material law operators. This setting essentially yields a new *normal form* for partial differential equations occurring in numerous applications.

In the following, we shall rigorously develop the solution theory of these abstract equations, which—due to their implied causality properties—we refer to as evolutionary equations. We use the term *evolutionary* in a somewhat subtle attempt to distinguish them from the classical concept of *evolution equations*, which are explicit first-order-in-time equations.

Although this class can be readily generalized to include more complicated cases, such as merely assuming that the numerical ranges of A, A^* are in the closed complex right half-plane or allowing for more complicated material law operators $M\left(\partial_0^{-1}\right)$ with the positive definiteness constraint that for some $c_0 \in\,]0, \infty[$ the numerical range of $\partial_0 M\left(\partial_0^{-1}\right) - c_0$ is in the closed complex right half-plane (for all sufficiently large $\varrho \in\,]0, \infty[$) (see again e.g. [59, 75]), we shall focus here on the more easily accessible pure differential case.

Eventually, we aim at a solution theory with easy to check assumptions that lead to well-posedness of a rather large class of partial differential equations. Indeed, we will see that well-posedness of an evolutionary equation boils down to proving a numerical range constraint for certain bounded operators only.

In Chap. 1, we develop the functional analytical setting and the basic solution theory. Chapter 2 illustrates the theory for a number of model problems from mathematical physics. This concludes the book's core material. Chapter 3 addresses some of the issues that may arise when comparing our approach with some alternative, possibly more mainstream ideas for dealing with problems of the same type. Two appendices complement the book's material by providing additional ideas for expanding on the applicability of the approach, Appendix A, and collecting some background material from functional analysis as a study resource, Appendix B.

Contents

The Solution Theory for a Basic Class of Evolutionary Equations

1.1 The Time Derivative

We start out with the definition of the time derivative. We emphasize that all vector spaces discussed in this exposition have the real numbers as underlying scalar field. This is a simplifying assumption. If given a complex Hilbert space, restrict the underlying scalar field to real multipliers and the scalar product to its real part. In this way the results developed here apply to the complex Hilbert space case as well. Note that, however, this reasoning can also be dispensed with and the complex Hilbert space case may be addresses directly, see the original work [58] for this. The exposition roughly follows [88, Chapter 1].

Definition 1.1.1 (Time Derivative) Let $L^2(\mathbb{R})$ be the Hilbert space of (equivalence classes of) square-integrable real-valued functions on $\mathbb{R}$. For $\varrho \in \mathbb{R}$ we define $H_{\varrho,0}(\mathbb{R}) := \{f \in L^2_{\text{loc}}(\mathbb{R}); (t \mapsto \exp(-\varrho t)f(t)) \in L^2(\mathbb{R})\}$ as a Hilbert space equipped with the inner product

$$\langle u|v\rangle_{\varrho,0} := \int_{\mathbb{R}} u(t)v(t)\exp(-2\varrho t)\,dt \quad (u,v \in H_{\varrho,0}(\mathbb{R})).$$

We set

$$\partial_{0,\varrho}|_{\mathring{C}_1(\mathbb{R})} : \mathring{C}_1(\mathbb{R}) \subseteq H_{\varrho,0}(\mathbb{R}) \to H_{\varrho,0}(\mathbb{R}),$$

$$u \mapsto u',$$

where $\mathring{C}_1(\mathbb{R})$ is the space of compactly supported continuously differentiable functions.

© Springer Nature Switzerland AG 2020
R. Picard et al., *A Primer for a Secret Shortcut to PDEs of Mathematical Physics*,
Frontiers in Mathematics, https://doi.org/10.1007/978-3-030-47333-4_1

Clearly, for all $\varrho \in \mathbb{R}$, the operator $\partial_{0,\varrho}|_{\mathring{C}_1(\mathbb{R})}$ is densely defined. The operator is also closable:

Proposition 1.1.2 *For all $v \in \mathring{C}_1(\mathbb{R})$ we have*

$$\left(\partial_{0,\varrho}|_{\mathring{C}_1(\mathbb{R})}\right)^* v = -v' + 2\varrho v,$$

hence $\partial_{0,\varrho}|_{\mathring{C}_1(\mathbb{R})}$ is closable.

Proof We note here that it suffices to prove the asserted equality. For if the equality is true, the adjoint of $\partial_{0,\varrho}|_{\mathring{C}_1(\mathbb{R})}$ is densely defined and thus $\partial_{0,\varrho}|_{\mathring{C}_1(\mathbb{R})}$ is closable. So, let $u, v \in \mathring{C}_1(\mathbb{R})$. Then we compute with the help of integration by parts

$$\langle \partial_{0,\varrho}|_{\mathring{C}_1(\mathbb{R})} u | v \rangle_{\varrho,0} = \langle u' | v \rangle_{\varrho,0}$$

$$= \int_{\mathbb{R}} u'(t) v(t) \exp(-2\varrho t) dt$$

$$= -\int_{\mathbb{R}} \left(u(t) v'(t) \exp(-2\varrho t) - 2\varrho u(t) v(t) \exp(-2\varrho t) \right) dt$$

$$= \langle u | -v' \rangle_{\varrho,0} + \langle u | 2\varrho v \rangle_{\varrho,0}.$$

This yields the assertion.   □

We define

$$\partial_{0,\varrho} := \overline{\partial_{0,\varrho}|_{\mathring{C}_1(\mathbb{R})}}.$$

A consequence of the latter proposition is

$$- \partial_{0,\varrho} + 2\varrho \subseteq \partial_{0,\varrho}^*. \tag{1.1.1}$$

Among other things we will show in the following that here equality is true. The strategy of the proof is to consider the inverse of $\partial_{0,\varrho}$ first. We define

$$L_\varrho^1(\mathbb{R}) := \{ h \in L_{\text{loc}}^1(\mathbb{R}); (t \mapsto \exp(-\varrho t) h(t)) \in L^1(\mathbb{R}) \}$$

for all $\varrho \in \mathbb{R}$ and recall Young's inequality.

Proposition 1.1.3 (Young's Inequality) *Let $\varrho \in \mathbb{R}$, $h \in L_{\varrho}^1(\mathbb{R})$, $f \in \overset{\circ}{C}_1(\mathbb{R})$. Then for all $t \in \mathbb{R}$*

$$h * f(t) := \int_{\mathbb{R}} h(t-s)f(s)ds$$

is well-defined and $t \mapsto h * f(t) \in H_{\varrho,0}(\mathbb{R})$ with

$$|h * f|_{\varrho,0} \le |h|_{L_{\varrho}^1}|f|_{\varrho,0}$$

holds. In particular, $h*$ extends to a bounded linear operator on $H_{\varrho,0}(\mathbb{R})$ with $\|h * \| \le |h|_{L_{\varrho}^1}$.

Proof Note that by a change of variables,

$$h * f(t) = \int_{\mathbb{R}} h(s)f(t-s)ds$$

for all $t \in \mathbb{R}$. This implies the existence of the integral (and even the continuity of $h * f$ by Lebesgue's dominated convergence theorem). Next, we estimate using the Cauchy–Schwarz inequality

$$\|h * f\|_{\varrho,0}^2 = \int_{\mathbb{R}} \left| \int_{\mathbb{R}} h(t-s)f(s)ds \right|^2 \exp(-2\varrho t)dt$$

$$\le \int_{\mathbb{R}} \left(\int_{\mathbb{R}} |h(t-s)| \exp(-\varrho(t-s)) |f(s)| \exp(-\varrho s)ds \right)^2 dt$$

$$= \int_{\mathbb{R}} \left(\int_{\mathbb{R}} (|h(t-s)| \exp(-\varrho(t-s)))^{1/2+1/2} |f(s)| \exp(-\varrho s)ds \right)^2 dt$$

$$\le \int_{\mathbb{R}} \left(\int_{\mathbb{R}} |h(t-s)| \exp(-\varrho(t-s))ds \times \right.$$

$$\left. \times \int_{\mathbb{R}} |h(t-s')| \exp(-\varrho(t-s')) |f(s')|^2 \exp(-2\varrho s')ds' \right) dt$$

$$= |h|_{L_{\varrho}^1} \int_{\mathbb{R}} \int_{\mathbb{R}} |h(t-s')| \exp(-\varrho(t-s'))dt \, |f(s')|^2 \exp(-2\varrho s')ds'$$

$$= |h|_{L_{\varrho}^1}^2 |f|_{\varrho,0}^2,$$

yielding the assertion. $\square$

A particular application of the latter estimate concerns the following two cases

$$h = \chi_{[0,\infty[} \in L^1_\varrho(\mathbb{R}) \qquad (1.1.2)$$

and

$$h = -\chi_{]-\infty,0]} \in L^1_{-\varrho}(\mathbb{R}) \qquad (1.1.3)$$

for all $\varrho > 0$. Note that in either case, we have $|h|_{L^1_\varrho} = 1/|\varrho|$ for $\varrho \neq 0$. Moreover, it is easy to see that $\chi_{[t,\infty)}(\cdot) f(\cdot) \in L^1_\varrho(\mathbb{R})$ for all $f \in H_{\varrho,0}(\mathbb{R})$, $t \in \mathbb{R}$, $\varrho > 0$, so that

$$t \mapsto h * f(t) = \int_{-\infty}^t f(s)ds$$

is well-defined and continuous (and analogously for $\varrho < 0$).

With these settings at hand, we prove the bounded invertibility of $\partial_{0,\varrho}$, $\varrho \neq 0$:

Theorem 1.1.4 *Let $\varrho \neq 0$. Then the operator $\partial_{0,\varrho}$ is continuously invertible, $\partial_{0,\varrho}^{-1} = h*$ with h respective of the sign of ϱ as in (1.1.2) or (1.1.3) and*

$$\|\partial_{0,\varrho}^{-1}\| \leq \frac{1}{|\varrho|}.$$

Proof We only prove the case $\varrho > 0$, the case $\varrho < 0$ being analogous. Let $f \in \mathring{C}_1(\mathbb{R})$ and let $\varphi \in \mathring{C}_1(\mathbb{R})$ be such that $0 \leq \varphi \leq 1$, $\varphi = 1$ on $[-1,1]$. For $n \in \mathbb{N}_{>0}$ we denote $\varphi_n := \varphi\left(\frac{\cdot}{n}\right)$. Then, by the fundamental theorem of calculus, we get

$$\partial_{0,\varrho}(\varphi_n(h * f)) = \partial_{0,\varrho}|_{\mathring{C}_1(\mathbb{R})} (\varphi_n(h * f))$$

$$= (\varphi_n(h * f))'$$

$$= \varphi_n'(h * f) + \varphi_n f$$

$$= \frac{1}{n}\varphi'\left(\frac{\cdot}{n}\right)(h * f) + \varphi_n f.$$

Letting $n \to \infty$, we deduce $h * f \in \text{dom}(\partial_{0,\varrho})$ and $\partial_{0,\varrho}(h * f) = f$. Indeed, this follows from $\varphi_n(h * f) \to h * f$ and $\frac{1}{n}\varphi'\left(\frac{\cdot}{n}\right)(h * f) + \varphi_n f \to f$ in $H_{\varrho,0}(\mathbb{R})$ and the closedness of $\partial_{0,\varrho}$. Next, for $f \in H_{\varrho,0}(\mathbb{R})$ there exists a sequence $(f_n)_n$ in $\mathring{C}_1(\mathbb{R})$ such that $f_n \to f$ in $H_{\varrho,0}(\mathbb{R})$. By Proposition 1.1.3, we deduce that $h * f_n \to h * f$ in $H_{\varrho,0}(\mathbb{R})$. And so, from $\partial_{0,\varrho}h * f_n = f_n$ we deduce that $h * f \in \text{dom}(\partial_{0,\varrho})$ and $\partial_{0,\varrho}(h * f) = f$.

Next, let $f \in \text{dom}(\partial_{0,\varrho})$ and $g := \partial_{0,\varrho}f$. There exists a sequence $(f_n)_n$ in $\mathring{C}_1(\mathbb{R})$ with the property that $f_n \to f$ and $g_n := \partial_{0,\varrho}f_n = f_n' \to g$ as $n \to \infty$ in $H_{\varrho,0}(\mathbb{R})$, by definition of $\partial_{0,\varrho}$. Thus, by Proposition 1.1.3 and the fundamental theorem of calculus

$$
\begin{aligned}
h * \partial_{0,\varrho} f &= h * g \\
&= \lim_{n\to\infty} h * g_n \\
&= \lim_{n\to\infty} h * f'_n \\
&= \lim_{n\to\infty} f_n = f,
\end{aligned}
$$

which yields the assertion. □

Corollary 1.1.5 *Let $\varrho \in \mathbb{R}$. Then*

$$
\partial_{0,\varrho}^* = -\partial_{0,\varrho} + 2\varrho.
$$

Proof Consider the unitary mapping

$$
\exp(-\varrho m) : H_{\varrho,0}(\mathbb{R}) \to L^2(\mathbb{R})
$$
$$
u \mapsto (t \mapsto \exp(-\varrho t)u(t))
$$

and its adjoint/inverse

$$
\exp(-\varrho m)^* : L^2(\mathbb{R}) \to H_{\varrho,0}(\mathbb{R})
$$
$$
v \mapsto (t \mapsto \exp(\varrho t)v(t)) .
$$

Then an easy computation shows

$$
\partial_{0,\varrho} = \exp(-\varrho m)^* \left(\partial_{0,0} + \varrho\right) \exp(-\varrho m). \tag{1.1.4}
$$

Indeed, the result is clear for elements in $\overset{\circ}{C}_1(\mathbb{R})$ and by taking closures, the equality follows. In particular, we see that the operators

$$
\partial_{0,0} \pm 1
$$

are boundedly invertible on $L^2(\mathbb{R})$ since both are unitarily equivalent to the invertible operators $\partial_{0,\pm 1}$, respectively. Since by (1.1.1) we have that

$$
\partial_{0,0} + 1 \subseteq -\partial_{0,0}^* + 1 = -(\partial_{0,0} - 1)^*
$$

we derive equality, because the operator on the left-hand side is onto and the operator on the right-hand side is one-to-one (using Theorem B.4.8). Summarizing, we have shown

$$\partial_{0,0} = -\partial_{0,0}^*.$$

According to (1.1.4) this, however, implies

$$\partial_{0,\varrho}^* = \exp(-\varrho m)^* \left(\partial_{0,0} + \varrho\right)^* \exp(-\varrho m)$$
$$= \exp(-\varrho m)^* \left(-\partial_{0,0} + \varrho\right) \exp(-\varrho m)$$
$$= -\partial_{0,\varrho} + 2\varrho,$$

which shows the claim.   □

We remark here that another consequence of the equality in Corollary 1.1.5 is that $\mathring{C}_1(\mathbb{R})$ is an operator core not only for $\partial_{0,\varrho}$ but also for $\partial_{0,\varrho}^*$. Moreover, we obtain that $\text{dom}(\partial_{0,\varrho}) = \text{dom}(\partial_{0,\varrho}^*)$. Note that the results obtained in this section carry over almost verbatim to the case of H-valued $H_{\varrho,0}$-functions, that is, to functions in the space

$$H_{\varrho,0}(\mathbb{R}; H) := \{f \in L^2_{\text{loc}}(\mathbb{R}; H); (t \mapsto \exp(-\varrho t)f(t)) \in L^2(\mathbb{R}; H)\}.$$

We summarize this in the following theorem, which for simplicity we only formulate for the case $\varrho > 0$.

Theorem 1.1.6 *Let $\varrho \in \mathbb{R}_{>0}$, and let H be a Hilbert space. Define*

$$\partial_{0,\varrho}|_{\mathring{C}_1(\mathbb{R};H)} : \mathring{C}_1(\mathbb{R}; H) \subseteq H_{\varrho,0}(\mathbb{R}; H) \to H_{\varrho,0}(\mathbb{R}; H), \varphi \mapsto \varphi'.$$

Then $\partial_{0,\varrho}|_{\mathring{C}_1(\mathbb{R};H)}$ is densely defined and closable; $\partial_{0,\varrho} := \overline{\partial_{0,\varrho}|_{\mathring{C}_1(\mathbb{R};H)}}$ is continuously invertible and for all $f \in H_{\varrho,0}(\mathbb{R}; H)$ we have

$$\partial_{0,\varrho}^{-1} f(t) = \int_{-\infty}^{t} f(s)\,ds \quad (t \in \mathbb{R}).$$

Furthermore, $\|\partial_{0,\varrho}^{-1}\| \leq 1/|\varrho|$ and $\partial_{0,\varrho}^* = -\partial_{0,\varrho} + 2\varrho.$

For $\varrho > 0$, the formula for the inverse of $\partial_{0,\varrho}$ reveals that the solution u of the equation $\partial_{0,\varrho} u = f$ up to a certain time $t \in \mathbb{R}$ does not depend on the behavior of f from $t \in \mathbb{R}$ onwards. This property is called causality and will be described by means of an estimate in the following theorem. The additional linear operator $M_0 \in \mathcal{B}(H)$ mentioned in the following statement can be thought of being the identity operator on H on a first read.

Moreover, when applied to elements in $H_{\varrho,0}(\mathbb{R}; H)$, the operator M_0 is to be understood in the point-wise sense, that is, $(M_0 u)(t) := M_0(u(t))$ for each $u \in H_{\varrho,0}(\mathbb{R}; H)$. We note that the inequality will play a crucial role in our analysis of (evolutionary) partial differential equations to follow.

Theorem 1.1.7 *Let $\varrho \in \mathbb{R}_{>0}$, let H be a Hilbert space, and let $0 \leq M_0 = M_0^* \in L(H)$. Then for all $u \in \mathrm{dom}(\partial_{0,\varrho})$ and $a \in \mathbb{R}$ we have*

$$\langle \partial_{0,\varrho} M_0 u | \chi_{]-\infty,a]} u \rangle_{\varrho,0} \geq \varrho \langle \chi_{]-\infty,a]} M_0 u | \chi_{]-\infty,a]} u \rangle_{\varrho,0}.$$

Proof By Theorem 1.1.6, it suffices to prove the inequality for $u \in \overset{\circ}{C}_1(\mathbb{R}; H)$. We compute for $a \in \mathbb{R}$ using integration by parts and the fact that $M_0 u \in \overset{\circ}{C}_1(\mathbb{R}; H)$, by the linearity, boundedness and selfadjointness of M_0,

$$\langle \partial_{0,\varrho} M_0 u | \chi_{]-\infty,a]} u \rangle_{\varrho,0}$$

$$= \int_{-\infty}^a \langle (M_0 u)'(s) | u(s) \rangle \exp(-2\varrho s) ds$$

$$= -\int_{-\infty}^a \langle M_0 u(s) | u'(s) \rangle \exp(-2\varrho s) ds$$

$$\quad + 2\varrho \int_{-\infty}^a \langle M_0 u(s) | u(s) \rangle \exp(-2\varrho s) ds + \langle M_0 u(a) | u(a) \rangle \exp(-2\varrho a)$$

$$\geq -\int_{-\infty}^a \langle u(s) | (M_0 u)'(s) \rangle \exp(-2\varrho s) ds + 2\varrho \int_{-\infty}^a \langle M_0 u(s) | u(s) \rangle \exp(-2\varrho s) ds.$$

Thus, we obtain

$$\langle \partial_{0,\varrho} M_0 u | \chi_{]-\infty,a]} u \rangle_{\varrho,0} + \langle \chi_{]-\infty,a]} u | \partial_{0,\varrho} M_0 u \rangle_{\varrho,0}$$

$$= 2 \langle \partial_{0,\varrho} M_0 u | \chi_{]-\infty,a]} u \rangle_{\varrho,0} \geq 2\varrho \langle \chi_{]-\infty,a]} M_0 u | \chi_{]-\infty,a]} u \rangle_{\varrho,0}. \qquad \square$$

To simplify notation we shall write ∂_0 instead of $\partial_{0,\varrho}$ if ϱ is clear from the context. Although in the one-dimensional case the index 0 is not really needed, we use this notation to underscore that ∂_0 will serve as our realization of the time derivative. (We anticipate the introduction of 'spatial' derivatives for which we shall use the indices starting with 1.)

A particular instance of Theorem 1.1.7 is $M_0 = 1$: Then we have

$$\langle \chi_{]-\infty,a]} u | \partial_0 u \rangle_{\varrho,0} = \langle u | \chi_{]-\infty,a]} \partial_0 u \rangle_{\varrho,0} \geq \varrho_0 \langle \chi_{]-\infty,a]} u | \chi_{]-\infty,a]} u \rangle_{\varrho,0}$$

for all $a \in \mathbb{R}$ and all $\varrho \in [\varrho_0, \infty[$, which precisely underpins the property of causality mentioned above: if $f = \partial_0 u$ vanishes on an interval $]-\infty, a]$ then so does $\partial_0^{-1} f = u$.

This property can also be expressed in the form

$$\chi_{]-\infty,a]}\,\partial_0^{-1}\left(1-\chi_{]-\infty,a]}\right)=0$$

or

$$\chi_{]-\infty,a]}\,\partial_0^{-1}=\chi_{]-\infty,a]}\,\partial_0^{-1}\chi_{]-\infty,a]}$$

for all $a \in \mathbb{R}$. Before we turn to partial differential equations, we will consider the derivative just defined in the context of ordinary differential equations, see also [22, 75]. The following corollary however, while involving only the one derivative, is essential for our analysis of partial differential equations.

Corollary 1.1.8 *Let H Hilbert space, $\varrho, \varepsilon > 0$. Then both $1 + \varepsilon\partial_0$ and $1 + \varepsilon\partial_0^*$ are continuously invertible. The operator norm of the inverses are bounded by 1 and*

$$(1 + \varepsilon\partial_0)^{-1},\ \left((1 + \varepsilon\partial_0)^{-1}\right)^* = \left(1 + \varepsilon\partial_0^*\right)^{-1} \to 1_{H_{\varrho,0}(\mathbb{R};H)}$$

in the strong operator topology as $\varepsilon \to 0$.

Proof Let $u \in \operatorname{dom}(\partial_0) = \operatorname{dom}(\partial_0^*)$ (see Theorem 1.1.6). We compute with the help of Theorem 1.1.7:

$$\langle(1 + \varepsilon\partial_0)\,u|u\rangle_{\varrho,0} = \langle u|\left(1 + \varepsilon\partial_0^*\right)u\rangle_{\varrho,0} \geq \langle u|u\rangle_{\varrho,0} + \varepsilon\varrho\langle u|u\rangle_{\varrho,0} \geq \langle u|u\rangle_{\varrho,0}.$$

Furthermore, from $(1 + \varepsilon\partial_0)^{-1}u = u - \varepsilon\,(1 + \varepsilon\partial_0)^{-1}\partial_0 u \to u$ as $\varepsilon \to 0$ for all $u \in \operatorname{dom}(\partial_0)$ and from $\sup_{\varepsilon>0}\|(1 + \varepsilon\partial_0)^{-1}\| \leq 1$, we deduce the first convergence statement. The second one is similar.   □

Remark 1.1.9 This corollary is a special case of Lemma B.7.1. Indeed, it suffices to observe that causality of $\partial_{0,\varrho}$ (in the form of Theorem 1.1.7) particularly implies the accretivity of $\partial_{0,\varrho}$ and of its adjoint $\partial_{0,\varrho}^*$.

1.2 A Hilbert Space Perspective on Ordinary Differential Equations

The above discussion suggests a Hilbert space theory for ordinary differential equations, which we explore for a moment. A more detailed exposition can be found in [22, 75] for the Hilbert space and [64] for the Banach space case.

Indeed, assuming henceforth the forward causal case of $\varrho \in \,]0, \infty[$, we have (see Theorem 1.1.4)[1]

$$\left\| \partial_0^{-1} \right\| \leq \frac{1}{\varrho}.$$

Remark 1.2.1 We note that the norm in $H_{\varrho,0}(\mathbb{R})$ is a Hilbert space variant of the Morgenstern norm, [36]. Based on the knowledge of the fundamental solution $h = \chi_{[0,\infty[}$ associated with ∂_0 we have on $L^\infty_{\text{loc}}(\mathbb{R})$-functions f with

$$\sup \{\exp(-\varrho t) \,|f(t)| \,;\, t \in \mathbb{R}\} < \infty,$$

that is, on $L^\infty_\varrho(\mathbb{R}) := \{f \in L^\infty_{\text{loc}}(\mathbb{R}); \sup\{\exp(-\varrho t)\,|f(t)| \,;\, t \in \mathbb{R}\} < \infty\}$, that

$$\partial_0^{-1} = \chi_{[0,\infty[} * .$$

We recall that by Theorem 1.1.4 the same formula is true in $H_{\varrho,0}(\mathbb{R})$. The continuity on $L^\infty_\varrho(\mathbb{R})$ can be confirmed easily by estimating

$$\left| \exp(-\varrho t) \int_{-\infty}^t f(s)\, ds \right| \leq \left| \exp(-\varrho t) \int_{-\infty}^t \exp(\varrho s) \exp(-\varrho s) \,|f(s)|\, ds \right|,$$

$$\leq \left| \exp(-\varrho t) \int_{-\infty}^t \exp(\varrho s)\, ds \right| \,|f|_{L^\infty_\varrho(\mathbb{R})} = \frac{1}{\varrho}\, |f|_{L^\infty_\varrho(\mathbb{R})}$$

for all $t \in \mathbb{R}$ and $f \in L^\infty_\varrho(\mathbb{R})$ and recalling that

$$\left(\partial_0^{-1} f \right)(t) = \int_\mathbb{R} \chi_{[0,\infty[}(t-s)\, f(s)\, ds$$

$$= \int_{-\infty}^t f(s)\, ds.$$

---

[1] Indeed, one can even confirm that $\left\| \partial_{0,\varrho}^{-1} \right\| = \frac{1}{\varrho}$.

Thus,

$$\left|\partial_0^{-1} f\right|_{L_\varrho^\infty(\mathbb{R})} \leq \frac{1}{\varrho} |f|_{L_\varrho^\infty(\mathbb{R})} \quad (f \in L_\varrho^\infty(\mathbb{R})).$$

Similarly, for the subspace

$$BC_\varrho([0, \infty[) := \{f \in L_\varrho^\infty(\mathbb{R}); \, \mathrm{supp}\, f \subseteq [0, \infty[, f|_{[0,\infty[} \text{ continuous}\} \subseteq L_\varrho^\infty(\mathbb{R}),$$

which is classically of particular interest (classical Morgenstern norm). Note that due to forward causality $\partial_0^{-1} \left[BC_\varrho([0, \infty[) \right] \subseteq BC_\varrho([0, \infty[).$

Next we present an ad-hoc application of Theorem 1.1.6, the description of the vector-valued version of the time derivative, to ordinary differential equations. For this we need the following notion:

Definition 1.2.2 Let $\varrho_0 \in \mathbb{R}$, H Hilbert space. Then we call

$$f: \; \mathrm{dom}(f) \subseteq \bigcap_{\varrho \geq \varrho_0} H_{\varrho,0}(\mathbb{R}; H) \to \bigcap_{\varrho \geq \varrho_0} H_{\varrho,0}(\mathbb{R}; H)$$

evolutionary at ϱ_0, if $\mathrm{dom}(f)$ is dense in $H_{\varrho,0}(\mathbb{R}; H)$ for all $\varrho \geq \varrho_0$ and if for all $\varrho \geq \varrho_0$ the mapping f extends to a Lipschitz continuous mapping $f_\varrho: H_{\varrho,0}(\mathbb{R}; H) \to H_{\varrho,0}(\mathbb{R}; H)$ with the property that

$$\sup_{\varrho \geq \varrho_0} |f_\varrho|_{\mathrm{Lip}} < \infty,$$

where $|f_\varrho|_{\mathrm{Lip}}$ denotes the Lipschitz semi-norm of f_ϱ. For evolutionary f, we always denote its Lipschitz continuous extension to $H_{\varrho,0}(\mathbb{R}; H)$ by f_ϱ.

Theorem 1.2.3 *Let* $\varrho_0 \in \mathbb{R}_{>0}$, *$H$ a Hilbert space and let,* $f: \mathrm{dom}(f) \subseteq \bigcap_{\varrho \geq \varrho_0} H_{\varrho,0}(\mathbb{R}; H) \to \bigcap_{\varrho \geq \varrho_0} H_{\varrho,0}(\mathbb{R}; H)$ *be evolutionary at* ϱ_0. *Then, for* $\varrho > \max\left\{\varrho_0, \sup_{\varrho \geq \varrho_0} |f_\varrho|_{\mathrm{Lip}}\right\}$ *there is a unique* $u_\varrho \in H_{\varrho,0}(\mathbb{R}; H)$ *satisfying*

$$\partial_{0,\varrho} u_\varrho = f_\varrho (u_\varrho).$$

Proof Define $L := \sup_{\varrho \geq \varrho_0} |f_\varrho|_{\mathrm{Lip}}$ and let $\varrho > \max\{\varrho_0, L\}$. Then

$$\left|\partial_0^{-1} f_\varrho(u) - \partial_0^{-1} f_\varrho(v)\right|_{\varrho,0} \leq \frac{1}{\varrho} |f_\varrho|_{\mathrm{Lip}} |u - v|_{\varrho,0}$$

$$\leq \frac{1}{\varrho} L |u - v|_{\varrho,0}.$$

Since $L/\varrho < 1$, $\partial_{0,\varrho}^{-1} f_\varrho$ is a contraction and by Banach's fixed point theorem existence and uniqueness of a $u_\varrho \in H_{\varrho,0}(\mathbb{R}; H)$ with

$$u_\varrho = \partial_{0,\varrho}^{-1} f_\varrho (u_\varrho)$$

follows. This is, in turn, equivalent to solving the above ODE. $\square$

As it stands the above theorem asserts something about the existence and uniqueness of a solution u_ϱ for every sufficiently large ϱ. In principle, however for different large enough parameters ϱ_1 and ϱ_2 we could have $u_{\varrho_1} \neq u_{\varrho_2}$. The reason for this is the fact that $f_{\varrho_1} \neq f_{\varrho_2}$ is possible also on $H_{\varrho_1}(\mathbb{R}; H) \cap H_{\varrho_2}(\mathbb{R}; H)$. Indeed, this follows from the observation that the norms on $H_{\varrho_1}(\mathbb{R}; H)$ and $H_{\varrho_2}(\mathbb{R}; H)$ cannot be compared. Interestingly, it turns out that one can avoid such an effect, if one assumes an additional requirement on f, namely that of causality. For this, we adopt the notion from the linear setting commented on at the end of the previous section.

Definition 1.2.4 Let $\varrho_0 \in \mathbb{R}$, H a Hilbert space. Let f be evolutionary at ϱ_0. We say that f is *causal* if, for all $u \in \mathrm{dom}(f)$ and $a \in \mathbb{R}$, we have $\chi_{]-\infty,a]}u \in \mathrm{dom}(f)$ and

$$\chi_{]-\infty,a]} f (u) = \chi_{]-\infty,a]} f \left(\chi_{]-\infty,a]} u \right). \tag{1.2.1}$$

Remark 1.2.5 If f is causal and $\varrho \geq \varrho_0$, then its Lipschitz continuous extension f_ϱ is causal, too. Indeed, since $\mathrm{dom}(f)$ is dense in $H_{\varrho,0}(\mathbb{R}; H)$ and f_ϱ and multiplication with $\chi_{]-\infty,a]}$ are continuous on $H_{\varrho,0}(\mathbb{R}; H)$, (1.2.1) follows for all $u \in H_{\varrho,0}(\mathbb{R}; H)$ and f_ϱ instead of f. We note here that to obtain causality for the continuous extension of a mapping (without the condition that multiplication by $\chi_{]-\infty,a]}$ leaves the domain invariant), knowing causality for the original mapping is subtle in general, see [86, 88].

Lemma 1.2.6 *Let $\varrho_0 \in \mathbb{R}$, H a Hilbert space, and f evolutionary at ϱ_0 and causal. Then, for $\varrho_1, \varrho_2 \in \mathbb{R}_{\geq \varrho_0}$ and $u \in H_{\varrho_1}(\mathbb{R}; H) \cap H_{\varrho_2}(\mathbb{R}; H)$ we have*

$$f_{\varrho_1} (u) = f_{\varrho_2}(u).$$

Proof Without loss of generality, we assume $\varrho_1 \leq \varrho_2$. Let $u \in H_{\varrho_1,0}(\mathbb{R}; H) \cap H_{\varrho_2,0}(\mathbb{R}; H)$. There exists $(u_k)_{k\in\mathbb{N}}$ in $\mathrm{dom}(f)$ with the property that $u_k \to u$ as $k \to \infty$ in $H_{\varrho_2,0}(\mathbb{R}; H)$. Let $a \in \mathbb{R}$. Then $\chi_{]-\infty,a]}u_k \to \chi_{]-\infty,a]}u$ as $k \to \infty$ in $H_{\varrho_1}(\mathbb{R}; H) \cap H_{\varrho_2}(\mathbb{R}; H)$. Indeed, this follows from the inequality

$$\left| \chi_{]-\infty,a]}v \right|_{\varrho_1,0}^2 = \int_{\mathbb{R}} \left| \chi_{]-\infty,a]} (t) \, v (t) \right|_H^2 \exp (-2\varrho_1 t) \, dt$$

$$= \int_{-\infty}^{a} |v (t)|_H^2 \exp (-2\varrho_1 t) \, dt$$

$$= \exp\left(-2\varrho_1 a\right) \int_{-\infty}^{a} |v\left(t\right)|_H^2 \exp\left(-2\varrho_1\left(t-a\right)\right) \, dt$$

$$\leq \exp\left(-2\varrho_1 a\right) \int_{-\infty}^{a} |v\left(t\right)|_H^2 \exp\left(-2\varrho_2\left(t-a\right)\right) \, dt$$

$$= \exp\left(-2\left(\varrho_1-\varrho_2\right)a\right) \left|\chi_{]-\infty,a]} v\right|_{\varrho_2,0}^2$$

$$\leq \exp\left(-2\left(\varrho_1-\varrho_2\right)a\right) |v|_{\varrho_2,0}^2$$

valid for all $v \in H_{\varrho_2,0}(\mathbb{R}; H)$. Thus, using Remark 1.2.5, we have for all $a \in \mathbb{R}$

$$\chi_{]-\infty,a]} f_{\varrho_1}(u) = \chi_{]-\infty,a]} f_{\varrho_1}(\chi_{]-\infty,a]} u) = \chi_{]-\infty,a]} \lim_{k \to \infty} f_{\varrho_1}(\chi_{]-\infty,a]} u_k)$$

$$= \chi_{]-\infty,a]} \lim_{k \to \infty} f(\chi_{]-\infty,a]} u_k) = \chi_{]-\infty,a]} \lim_{k \to \infty} f_{\varrho_2}(\chi_{]-\infty,a]} u_k) = \chi_{]-\infty,a]} f_{\varrho_2}(u).$$

Hence, the claim follows.   □

The anticipated theorem that the solution is independent of the parameter ϱ reads as follows:

Theorem 1.2.7 *Let $\varrho_0 \in \mathbb{R}_{>0}$, H Hilbert space. Assume f to be evolutionary at ϱ_0 and causal. Let $\varrho_1, \varrho_2 \in \mathbb{R}_{>\max\{\varrho_0, L\}}$, where $L := \sup_{\varrho \geq \varrho_0} |f_\varrho|_{\mathrm{Lip}}$, and let $u_{\varrho_k} \in H_{\varrho_k,0}(\mathbb{R}; H)$ with*

$$u_{\varrho_k} = \partial_{0,\varrho_k}^{-1} f_{\varrho_k}\left(u_{\varrho_k}\right)$$

for $k \in \{1, 2\}$. Then

$$u_{\varrho_1} = u_{\varrho_2} \in H_{\varrho_1,0}\left(\mathbb{R}; H\right) \cap H_{\varrho_2,0}\left(\mathbb{R}; H\right).$$

Proof Without loss of generality, we assume that $\varrho_1 \leq \varrho_2$. We consider the mapping

$$\chi_{[0,\infty[} * f(\cdot) : \mathrm{dom}(f) \subseteq \bigcap_{\varrho \geq \varrho_0} H_{\varrho,0}(\mathbb{R}; H) \to \bigcap_{\varrho \geq \varrho_0} H_{\varrho,0}(\mathbb{R}; H),$$

which is causal and evolutionary at ϱ_0. Its unique Lipschitz continuous extension on $H_{\varrho,0}(\mathbb{R}; H)$ is given by $\partial_{0,\varrho}^{-1} f_\varrho$ for $\varrho \geq \varrho_0$, by Theorem 1.1.7. Let $a \in \mathbb{R}$. We note that due to causality, we have that

$$\chi_{]-\infty,a]} u_{\varrho_1} = \chi_{]-\infty,a]} \partial_{0,\varrho_1}^{-1} f_{\varrho_1}(u_{\varrho_1}) = \chi_{]-\infty,a]} \partial_{0,\varrho_1}^{-1} f_{\varrho_1}(\chi_{]-\infty,a]} u_{\varrho_1}),$$

that is, $\chi_{]-\infty,a]}u_{\varrho_1}$ is a fixed point of $\chi_{]-\infty,a]}\partial_{0,\varrho_1}^{-1}f_{\varrho_1}$, which is unique due to $|\chi_{]-\infty,a]}\partial_{0,\varrho_1}^{-1}f_{\varrho_1}|_{\text{Lip}} < 1$. On the other hand, we have that $\chi_{]-\infty,a]}u_{\varrho_2} \in H_{\varrho_1}(\mathbb{R}; H) \cap H_{\varrho_2}(\mathbb{R}; H)$ and thus, by Lemma 1.2.6,

$$
\begin{aligned}
\chi_{]-\infty,a]}u_{\varrho_2} &= \chi_{]-\infty,a]}\partial_{0,\varrho_2}^{-1}f_{\varrho_2}(u_{\varrho_2}) \\
&= \chi_{]-\infty,a]}\partial_{0,\varrho_2}^{-1}f_{\varrho_2}(\chi_{]-\infty,a]}u_{\varrho_2}) \\
&= \chi_{]-\infty,a]}\partial_{0,\varrho_1}^{-1}f_{\varrho_1}(\chi_{]-\infty,a]}u_{\varrho_2}),
\end{aligned}
$$

that is, $\chi_{]-\infty,a]}u_{\varrho_2}$ is also a fixed point of $\chi_{]-\infty,a]}\partial_{0,\varrho_1}^{-1}f_{\varrho_1}$ and hence,

$$
\chi_{]-\infty,a]}u_{\varrho_1} = \chi_{]-\infty,a]}u_{\varrho_2}.
$$

Since $a \in \mathbb{R}$ was arbitrary, the result follows. $\qquad\square$

The type of equations considered in these theorems include differential equations with delay, see [22, 64, 75] for a more in-depth discussion.

We conclude our discussion of ordinary differential equations by considering initial value problems in view of Theorems 1.2.3 and 1.2.7:

Remark 1.2.8 Let $\varrho_0 \in \mathbb{R}$, and let H be a Hilbert space. Let f be evolutionary at ϱ_0 and causal, $u_0 \in H$ and $\varrho \in \mathbb{R}_{>\max\{\varrho_0, L\}}$ with L as in Theorem 1.2.7. An initial condition can be written as

$$
\partial_{0,\varrho}\left(u - \chi_{]0,\infty[} \otimes u_0\right) = f_\varrho(u), \tag{1.2.2}
$$

where u_0 is the initial value at time 0. Here $\chi_{]0,\infty[} \otimes u_0$ is defined by

$$
\left(\chi_{]0,\infty[} \otimes u_0\right)(t) = \chi_{]0,\infty[}(t)\, u_0, \; t \in \mathbb{R}.
$$

Substituting $w := u - \chi_{]0,\infty[} \otimes u_0$, we see that solving the latter equation amounts to looking for a solution w of

$$
\partial_{0,\varrho}w = \widetilde{f_\varrho}(w),
$$

where

$$
\widetilde{f_\varrho}(w) := \chi_{]0,\infty[}f_\varrho\left(w + \chi_{]0,\infty[} \otimes u_0\right).
$$

Note that $\widetilde{f_\varrho}$ inherits the required Lipschitz property from f_ϱ. The desired solution u is then given by

$$
u := \chi_{]0,\infty[} \otimes u_0 + w.
$$

Causality of $\partial_{0,\varrho}^{-1}$ yields that w and so u must vanish on $]-\infty, 0[$. So we see that initial value problems were indeed already incorporated in the above, although in a slightly different perspective.

Now, we return to our discussion of partial differential equations.

1.3   Evolutionary Equations

In this section, we shall discuss the theoretic foundation of the problem class we have in mind to describe and solve partial differential equations (involving the 'time derivative' ∂_0). Having studied the time derivative and its properties in the previous section, we shall now exploit the (strict) accretivity of ∂_0 in more involved contexts. In fact, strict accretivity plays a crucial role when discussing (abstract) partial differential equations as sums of two unbounded operators in space-time.

The reader may roughly compare the present approach to the seminal papers [7, 10], where operator sums typically of the form $\frac{d}{dt} + A$ were considered. In the cases to be considered, we complement the problem class outlined in [7, 10] by considering (possibly non-regular) coefficients of $\frac{d}{dt}$ and restricting ourselves to a genuine Hilbert space setting. Furthermore, treating partial differential equations predominantly as first order systems, the approach to be described certainly has some flavor of symmetric hyperbolic systems as introduced in [16], see also [47]. Indeed, we might refer to our problem class as abstract Friedrichs systems.

1.3.1   The Problem Class

Throughout this section, we make the following standing assumptions. H is a real Hilbert space and $M_0, M_1 \in B(H)$. M_0 is selfadjoint and there exists $c_0 \in \mathbb{R}_{>0}$ and $\varrho_0 \in \mathbb{R}_{>0}$ such that for all $\varrho \in \mathbb{R}_{\geq \varrho_0}$

$$\langle \varrho M_0 \varphi | \varphi \rangle_H + \langle M_1 \varphi | \varphi \rangle_H \geq c_0 \langle \varphi | \varphi \rangle_H \quad (\varphi \in H). \tag{1.3.1}$$

Furthermore, $A\colon \operatorname{dom}(A) \subseteq H \to H$ is a skew-selfadjoint operator, that is[2] $A = -A^*$. Consequently, we have

$$2\langle A\varphi | \varphi \rangle_H = \langle A\varphi | \varphi \rangle_H + \langle \varphi | A\varphi \rangle_H$$
$$= \langle A\varphi | \varphi \rangle_H + \langle A^* \varphi | \varphi \rangle_H = \langle A\varphi | \varphi \rangle_H - \langle A\varphi | \varphi \rangle_H = 0 \tag{1.3.2}$$

---

[2]The skew-selfadjointness of A is responsible for an "energy balance law" without spatial derivatives. If

$$(\partial_0 M_0 + M_1 + A)u = f$$

for all $\varphi \in \text{dom}(A) = \text{dom}(A^*)$. In the following, we consider M_0, M_1 and A as operators extended to $H_{\varrho,0}(\mathbb{R}; H)$. We gather some results needed in the following.

Proposition 1.3.1 *Let $\varrho \in \mathbb{R}$ and let $C \colon \text{dom}(C) \subseteq H \to H$ be a densely defined, closed, linear operator and $T \in \mathcal{B}(H_{\varrho,0}(\mathbb{R}))$. Then:*

(1) *The operator $\widehat{C} \colon H_{\varrho,0}(\mathbb{R}; \text{dom}(C)) \subseteq H_{\varrho,0}(\mathbb{R}; H) \to H_{\varrho,0}(\mathbb{R}; H)$ with*

$$(\widehat{C}f)(t) := C(f(t))$$

for almost all $t \in \mathbb{R}$, $f \in \text{dom}(\widehat{C})$ is closed.
(2) *The set $\mathring{C}_1(\mathbb{R}) \cdot \text{dom}(C)$, defined as $\text{span}\{f(\cdot)x \in \mathring{C}_1(\mathbb{R}; \text{dom}(C)); f \in \mathring{C}_1(\mathbb{R}), x \in \text{dom}(C)\}$ is dense in $H_{\varrho,0}(\mathbb{R}; H)$ and indeed an operator core for $\widehat{C}$.*
(3) $\left(\widehat{C}\right)^* = \widehat{C^*}$.
(4) *Define $\widehat{T} \in \mathcal{B}(H_{\varrho,0}(\mathbb{R}; H))$ by linear and continuous extension of $\widehat{T}f(\cdot)x := (t \mapsto (Tf)(t)x)$ for all $f \in \mathring{C}_1(\mathbb{R})$ and $x \in H$. Then $\widehat{T}\widehat{C} \subseteq \widehat{C}\widehat{T}$.*

Proof

(1) Let $(f_n)_{n\in\mathbb{N}}$ be a sequence in $H_{\varrho,0}(\mathbb{R}; \text{dom}(C))$ and $f \in H_{\varrho,0}(\mathbb{R}; H)$ with the property that $f_n \to f$ and $\widehat{C}f_n \to g$ in $H_{\varrho,0}(\mathbb{R}; H)$ as $n \to \infty$ for some $g \in H_{\varrho,0}(\mathbb{R}; H)$. Without loss of generality, we may assume that there is a set $N \subseteq \mathbb{R}$ of Lebesgue measure zero with the property $f_n(t) \in \text{dom}(C)$ and $f_n(t) \to f(t)$ and $Cf_n(t) = (\widehat{C}f_n)(t) \to g(t)$ for all $t \in \mathbb{R} \setminus N$. Then, by the closedness of C, we conclude $f(t) \in \text{dom}(C)$ and $Cf(t) = g(t)$ for all $t \in \mathbb{R} \setminus N$. Thus, $f \in \text{dom}(\widehat{C})$ and $\widehat{C}f = g$.

---

then, since $\langle u | Au \rangle_H = 0$, we have (at least formally)

$$\frac{1}{2}\partial_0 \langle u | M_0 u \rangle_H + \langle u | M_1 u \rangle_H = \langle u | \partial_0 M_0 u \rangle_H + \langle u | M_1 u \rangle_H = \langle u | f \rangle_H .$$

By integration over a non-empty time-interval $[\tau, T]$, where f vanishes, we get

$$\frac{1}{2} \langle u | M_0 u \rangle_H (T) + \int_\tau^T \langle u | M_1 u \rangle_H = \frac{1}{2} \langle u | M_0 u \rangle_H (\tau) ,$$

which is an "energy balance law" in comparison to the initial energy $\frac{1}{2} \langle u | M_0 u \rangle_H (\tau)$ at time τ. If $M_1 = -M_1^*$, we have indeed conservation of "energy"

$$\frac{1}{2} \langle u | M_0 u \rangle_H (T) = \frac{1}{2} \langle u | M_0 u \rangle_H (\tau) .$$

We have set phrases like "energy balance" and "energy" in inverted commas, since from a mathematical perspective to introduce the concept of energy is inappropriate and unnecessary. Following common practice, we use those terms merely as jargon.

(2) Let $g \in \left(\mathring{C}_1(\mathbb{R}) \cdot \mathrm{dom}(C) \right)^\perp$, where the orthogonal complement is computed in $H_{\varrho,0}(\mathbb{R}; H)$. Since g is measurable, there exists a set $N_1 \subseteq \mathbb{R}$ of measure zero with $\overline{\mathrm{span}\, g[\mathbb{R} \setminus N_1]} =: H_0 \subseteq H$ separable.[3] In particular, since H_0 is separable, we find $(x_n)_{n \in \mathbb{N}}$ in $\mathrm{dom}(C)$ with the property that $\overline{\mathrm{span}\{x_n; n \in \mathbb{N}\}} \supseteq H_0$. Then, for all $\varphi \in \mathring{C}_1(\mathbb{R})$ and $n \in \mathbb{N}$ we deduce

$$0 = \langle g | \varphi x_n \rangle_{\varrho,0} = \int_{\mathbb{R}} \langle g(t) | \varphi(t) x_n \rangle_H \exp(-2\varrho t) dt,$$

$$= \int_{\mathbb{R}} \langle g(t) | x_n \rangle \exp(-2\varrho t) \varphi(t) dt.$$

Hence, there is a set $N_1 \subseteq N \subseteq \mathbb{R}$ of measure zero with the property that

$$\langle g(t) | x_n \rangle = 0 \quad (t \in \mathbb{R} \setminus N, \ n \in \mathbb{N}).$$

Thus, $g(t) = 0$ for all $t \in \mathbb{R} \setminus N$ and so $g = 0$. The statement on the operator core follows from the observation that simple $\mathrm{dom}(C)$-valued functions can be approximated by elements in $\mathring{C}_1(\mathbb{R}) \cdot \mathrm{dom}(C)$. Simple $\mathrm{dom}(C)$-valued functions are dense in $H_{\varrho,0}(\mathbb{R}; \mathrm{dom}(C))$. Hence, this assertion follows.

(3) Let $f \in \mathrm{dom}(\widehat{C^*})$ and $g \in \mathrm{dom}(\widehat{C})$. Then,

$$\langle \widehat{C} g | f \rangle_{\varrho,0} = \int_{\mathbb{R}} \langle C g(t) | f(t) \rangle \exp(-2\varrho t) \, dt$$

$$= \int_{\mathbb{R}} \langle g(t) | C^* f(t) \rangle \exp(-2\varrho t) \, dt$$

$$= \langle g | \widehat{C^*} f \rangle_{\varrho,0},$$

which shows $\widehat{C^*} \subseteq \left(\widehat{C} \right)^*$. On the other hand, let $f \in \mathrm{dom}\left(\left(\widehat{C} \right)^* \right)$, $\varphi \in \mathring{C}_1(\mathbb{R})$, and $x \in \mathrm{dom}(C)$. Then

$$\int_{\mathbb{R}} \varphi(t) \langle C x | f(t) \rangle \exp(-2\varrho t) \, dt = \langle \widehat{C} (\varphi(\cdot) x) | f \rangle_{\varrho,0}$$

$$= \left\langle \varphi(\cdot) x \, \middle| \left(\widehat{C} \right)^* f \right\rangle_{\varrho,0}$$

$$= \int_{\mathbb{R}} \varphi(t) \left\langle x \, \middle| \left(\left(\widehat{C} \right)^* f \right)(t) \right\rangle \exp(-2\varrho t) \, dt.$$

---

[3]Note that g is measurable if there is a sequence of simple function $g_n : \mathbb{R} \to H$ such that $g_n(t) \to g(t)$ for almost all $t \in \mathbb{R}$. Since g_n attains just finitely many values, we deduce that g is almost everywhere separably valued.

Since $\varphi \in \overset{\circ}{C}_1(\mathbb{R})$ was chosen arbitrarily, it follows that

$$\langle Cx | f(t) \rangle = \left\langle x \left| \left(\left(\widehat{C} \right)^* f \right)(t) \right. \right\rangle$$

for almost all $t \in \mathbb{R}$ and $x \in \text{dom}(C)$. The latter gives $f(t) \in \text{dom}(C^*)$ and $C^* f(t) = \left(\left(\widehat{C} \right)^* f \right)(t)$ for almost all $t \in \mathbb{R}$, which implies $f \in \text{dom}(\widehat{C^*})$.

(4) We note that $\widehat{T}$ is a well-defined, continuous linear operator with $\|\widehat{T}\| \leq \|T\|$. Then, for $f(\cdot)x \in \overset{\circ}{C}_1(\mathbb{R}) \cdot \text{dom}(C)$,

$$\begin{aligned}
\widehat{T}\left(\widehat{C} f(\cdot)x \right) &= \widehat{T}\left(\widehat{C}(t \mapsto f(t)x) \right) \\
&= \widehat{T}\left(t \mapsto Cf(t)x \right) \\
&= \widehat{T}\left(t \mapsto f(t)Cx \right) \\
&= (t \mapsto (Tf)(t)Cx) \\
&= (t \mapsto C(Tf)(t)x) \\
&= \widehat{C}\left(t \mapsto (Tf)(t)x \right) = \widehat{C}\widehat{T} f(\cdot)x.
\end{aligned}$$

Hence, $\widehat{C}\widehat{T} = \widehat{T}\widehat{C}$ on $\overset{\circ}{C}_1(\mathbb{R}) \cdot \text{dom}(C)$. Thus, by (2) and the closedness of $\widehat{C}$, (i.e., (1)), we obtain $\widehat{C}\widehat{T} \supseteq \widehat{T}\widehat{C}$. $\qquad\square$

In order to avoid unnecessarily cluttered notation – similarly to what we did for the time derivative – we shall simply re-use C for denoting $\widehat{C}$.

The problem we study in the following then takes the form

$$(\partial_0 M_0 + M_1 + A) U = F \tag{1.3.3}$$

in $H_{\varrho,0}(\mathbb{R}; H)$ for $\varrho \in \mathbb{R}_{\geq \varrho_0}$ (with ϱ_0 as in (1.3.1)).

Equations of the form in (1.3.3) are referred to as *evolutionary equations*.

Before we proceed to state and prove a well-posedness theorem for the latter equation, we note a subtlety. Note that the operator ∂_0 and maybe also A are unbounded operators. Thus, in (1.3.3) we may have to deal with the sum of two unbounded operators. A priori, this sum is defined on the intersection of the respective domains only, that is, on $\text{dom}(\partial_0 M_0) \cap \text{dom}(A)$. A solution theory for (1.3.3) would amount to showing that the operator $\partial_0 M_0 + M_1 + A$ is onto (Existence of solutions), one-to-one (Uniqueness of solutions) and has a bounded inverse (continuous dependence on the data). In general, this cannot be expected to be true since $\partial_0 M_0 + M_1 + A$ is not closed but only closable. This issue is conveniently by-passed by looking instead for a solution theory for

$$\overline{(\partial_0 M_0 + M_1 + A)} U = F, \tag{1.3.4}$$

where the closure is taken in $H_{\varrho,0}(\mathbb{R}; H)$. Then as a post-processing procedure, that is, with some regularity theory added, it is possible to show for particular M_0, M_1 and A that the operator sum $\partial_0 M_0 + M_1 + A$ is closed anyway, see, e.g., [71]. Later on, we shall use $\partial_0 M_0 + M_1 + A$ also as a simplified notation[4] for the closure $\overline{\partial_0 M_0 + M_1 + A}$ of $\partial_0 M_0 + M_1 + A$. We address the solution theory for (1.3.4) in the following.

1.3.2 The Solution Theory for Simple Material Laws

We continue in the setting described at the beginning of the previous section. So, let $M_0 = M_0^*$, $M_1 \in \mathcal{B}(H)$ satisfying (1.3.1) and let $A = -A^*$ in a Hilbert space H. The well-posedness theorem reads as follows:

Theorem 1.3.2 (Solution Theory and Causality) *Given $\varrho_0, c_0 \in \mathbb{R}_{>0}$ as in (1.3.1), for each $\varrho \geq \varrho_0$, the operator*

$$B: \operatorname{dom}(\partial_0) \cap \operatorname{dom}(A) \subseteq H_{\varrho,0}(\mathbb{R}; H) \to H_{\varrho,0}(\mathbb{R}; H)$$

$$U \mapsto (\partial_0 M_0 + M_1 + A)\, U$$

is densely defined and closable; $\overline{B}$ is continuously invertible, $\left\| \left(\overline{B}\right)^{-1} \right\| \leq 1/c_0$ and $\left(\overline{B}\right)^{-1}$ is causal, that is,

$$\chi_{]-\infty,a]} \left(\overline{B}\right)^{-1} \chi_{]-\infty,a]} = \chi_{]-\infty,a]} \left(\overline{B}\right)^{-1} \quad (a \in \mathbb{R}).$$

The proof of the solution theory is based on coercivity estimates for both B and B^*. The causality also follows from a refined coercivity estimate for B. This estimate is based on the following observation:

Proposition 1.3.3 *Let $\varrho_0, c_0 \in \mathbb{R}_{>0}$ as in (1.3.1) and $\varrho \geq \varrho_0$. Then for all $\varphi \in \operatorname{dom}(\partial_0) \cap \operatorname{dom}(A)$ and $a \in \mathbb{R}$ we have*

$$\langle (\partial_0 M_0 + M_1 + A)\, \varphi | \chi_{]-\infty,a]}\varphi \rangle_{\varrho,0} \geq c_0 \langle \chi_{]-\infty,a]}\varphi | \chi_{]-\infty,a]}\varphi \rangle_{\varrho,0}.$$

---

[4]There is a deeper reason using this notational convenience. If the application of ∂_0 and A is generalized in a "distributional sense" by re-using the notation ∂_0 for the dual of its adjoint $\left(\partial_0^*\right)^\diamond$ and A for the negative of its dual $A^\diamond$ (this notation of the dual assumes implicitly that we identify $H_{\varrho,0}(\mathbb{R}, H)$ with its dual; see also Proposition 2.4.3), we have indeed $\overline{M_0 \partial_0 + M_1 + A} \subseteq M_0 \partial_0 + M_1 + A \subseteq \partial_0 M_0 + M_1 + A$. To keep matters simple, we will not consider this generalization process further, compare, however, [58, 62, Chapter 6].

Proof By (1.3.2) and Proposition 1.3.1 (applied to $C = A$), we have for all $\varphi \in \text{dom}(A)$ and $a \in \mathbb{R}$:

$$\langle A\varphi | \chi_{]-\infty,a]}\varphi \rangle_{\varrho,0} = \int_{-\infty}^{a} \langle A\varphi(t) | \varphi(t) \rangle \exp(-2\varrho t) dt = 0.$$

Next, again by Proposition 1.3.1 (applied to $T = \chi_{]-\infty,a]}$ and $M_1 = C$), we obtain

$$\langle M_1 \varphi | \chi_{]-\infty,a]}\varphi \rangle_{\varrho,0} = \langle M_1 \chi_{]-\infty,a]}\varphi | \chi_{]-\infty,a]}\varphi \rangle_{\varrho,0} \quad (a \in \mathbb{R}).$$

These two equalities in conjunction with Theorem 1.1.7 yield for all $\varphi \in \text{dom}(\partial_0) \cap \text{dom}(A)$

$$\langle (\partial_0 M_0 + M_1 + A)\, \varphi | \chi_{]-\infty,a]}\varphi \rangle_{\varrho,0}$$
$$\geq \langle (\varrho M_0 + M_1)\, \chi_{]-\infty,a]}\varphi | \chi_{]-\infty,a]}\varphi \rangle_{\varrho,0} \geq c_0 \langle \chi_{]-\infty,a]}\varphi | \chi_{]-\infty,a]}\varphi \rangle_{\varrho,0},$$

where in the last step we used inequality (1.3.1) point-wise under the integral. $\qquad\square$

Remark 1.3.4 A closer look at the last proof reveals that for all $\varphi \in \text{dom}(\partial_0) \cap \text{dom}(A)$ and $a \in \mathbb{R}$ we obtain

$$\langle (\partial_0 M_0 + M_1 + A)\, \varphi, \chi_{]-\infty,a]}\varphi \rangle_{\varrho,0} \geq \langle (\varrho M_0 + M_1)\, \chi_{]-\infty,a]}\varphi, \chi_{]-\infty,a]}\varphi \rangle_{\varrho,0}.$$

Thus, letting $a \to \infty$, and using the fact that $\text{dom}(\partial_0) \cap \text{dom}(A)$ is (by definition) an operator core for B in Theorem 1.3.2, we obtain

$$\langle \overline{B}\varphi | \varphi \rangle_{\varrho,0} \geq \langle (\varrho M_0 + M_1)\, \varphi | \varphi \rangle_{\varrho,0} \quad (\varphi \in \text{dom}(\overline{B})). \tag{1.3.5}$$

Next, we compute the adjoint of B given in Theorem 1.3.2.

Theorem 1.3.5 *Let M_0, M_1, A, and B be as in Theorem 1.3.2. Then*

$$(\partial_0 M_0 + M_1 + A)^* = \overline{(M_0 \partial_0^* + M_1^* - A)}\,|_{\text{dom}(\partial_0^*) \cap \text{dom}(A)}.$$

Proof Let $u \in \text{dom}\left((\partial_0 M_0 + M_1 + A)^*\right)$. For $\varepsilon > 0$ we define $u_\varepsilon := \left(1 + \varepsilon \partial_0^*\right)^{-1} u$, which is well-defined by Corollary 1.1.8. For $v \in \text{dom}(\partial_0) \cap \text{dom}(A) = \text{dom}(B)$ we compute with the help of Proposition 1.3.1 (applied to $T = (1 + \varepsilon \partial_0)^{-1}$, $C \in \{M_0, M_1, A\}$):

$$\langle u_\varepsilon | (\partial_0 M_0 + M_1 + A)\, v \rangle_{\varrho,0} = \left\langle \left(1 + \varepsilon \partial_0^*\right)^{-1} u \,\middle|\, (\partial_0 M_0 + M_1 + A)\, v \right\rangle_{\varrho,0}$$

$$= \left\langle u \,\middle|\, (1 + \varepsilon \partial_0)^{-1} (\partial_0 M_0 + M_1 + A)\, v \right\rangle_{\varrho,0}$$

$$= \left\langle u \,|\, (\partial_0 M_0 + M_1 + A)\,(1 + \varepsilon \partial_0)^{-1}\, v \right\rangle_{\varrho,0}$$

$$= \left\langle \left(1 + \varepsilon \partial_0^*\right)^{-1} (\partial_0 M_0 + M_1 + A)^*\, u \,|\, v \right\rangle_{\varrho,0}. \tag{1.3.6}$$

Thus, $u_\varepsilon \in \mathrm{dom}(B^*)$ and

$$B^* u_\varepsilon = \left(1 + \varepsilon \partial_0^*\right)^{-1} (\partial_0 M_0 + M_1 + A)^*\, u = \left(1 + \varepsilon \partial_0^*\right)^{-1} B^* u. \tag{1.3.7}$$

Revisiting (1.3.6), we also obtain for all $v \in \mathrm{dom}(\partial_0) \cap \mathrm{dom}(A)$, which is dense in $\mathrm{dom}(A)$ by Proposition 1.3.1(2), that

$$\langle u_\varepsilon \,|\, A v \rangle_{\varrho,0} = \langle \left(1 + \varepsilon \partial_0^*\right)^{-1} B^* u - M_0 \partial_0^* u_\varepsilon - M_1^* u_\varepsilon \,|\, v \rangle_{\varrho,0},$$

where we have also used that $u_\varepsilon \in \mathrm{dom}(\partial_0^*)$. Thus, $u_\varepsilon \in \mathrm{dom}(A^*) = \mathrm{dom}(-A)$ and

$$-A u_\varepsilon = \left(1 + \varepsilon \partial_0^*\right)^{-1} B^* u - M_0 \partial_0^* u_\varepsilon - M_1^* u_\varepsilon.$$

This together with (1.3.7) yields

$$B^* u_\varepsilon = \left(1 + \varepsilon \partial_0^*\right)^{-1} B^* u = \left(M_0 \partial_0^* + M_1^* - A\right) u_\varepsilon. \tag{1.3.8}$$

Since $C_0 := \left(M_0 \partial_0^* + M_1^* - A\right)|_{\mathrm{dom}(\partial_0) \cap \mathrm{dom}(A)} \subseteq B^*$ and B is densely defined (see Proposition 1.3.1), we obtain that C_0 is closable; let $C := \overline{C_0}$. Now, let $\varepsilon \to 0$ in (1.3.8). Then $u_\varepsilon \to u$ and $\left(1 + \varepsilon \partial_0^*\right)^{-1} B^* u \to B^* u$ in $H_{\varrho,0}(\mathbb{R}; H)$ by Corollary 1.1.8. Hence, $u \in \mathrm{dom}\,(C)$ and $C u = B^* u$, which implies

$$C_0 \subseteq B^* \subseteq C.$$

Thus, $C = B^*$.   □

In terms of the simplifying notation we have adopted, we may write this result as

$$(\partial_0 M_0 + M_1 + A)^* = \partial_0^* M_0 + M_1^* - A.$$

Proof of Theorem 1.3.2 For $U \in \mathrm{dom}(\partial_0) \cap \mathrm{dom}(A)$ and all $a \in \mathbb{R}$ we have by Proposition 1.3.3

$$\langle B U \,|\, \chi_{]-\infty,a]} U \rangle_{\varrho,0} = \langle (\partial_0 M_0 + M_1 + A)\, U \,|\, \chi_{]-\infty,a]} U \rangle_{\varrho,0} \tag{1.3.9}$$

$$\geq c_0 \langle \chi_{]-\infty,a]} U \,|\, \chi_{]-\infty,a]} U \rangle_{\varrho,0}.$$

Letting $a \to \infty$, we obtain

$$\langle U \,|\, \big(M_0 \partial_0^* + M_1^* - A\big) U \rangle_{\varrho,0} = \langle (\partial_0 M_0 + M_1 + A) U \,|\, U \rangle_{\varrho,0} \geq c_0 \langle U | U \rangle_{\varrho,0}.$$

Thus,

$$\langle \overline{(\partial_0 M_0 + M_1 + A)} U | U \rangle_{\varrho,0} \geq c_0 \langle U | U \rangle_{\varrho,0} \quad \big(U \in \mathrm{dom}\,\big(\overline{(\partial_0 M_0 + M_1 + A)}\big)\big)$$

and

$$\langle V \,|\, (\partial_0 M_0 + M_1 + A)^* \, V \rangle_{\varrho,0} \geq c_0 \langle V | V \rangle_{\varrho,0} \quad \big(V \in \mathrm{dom}\,\big((\partial_0 M_0 + M_1 + A)^*\big)\big),$$

where in the last step we have used Theorem 1.3.5. Thus, the result on continuous invertibility of $\overline{B}$ now follows from (the elementary) Proposition B.6.3 of Appendix B. For the causality of $\overline{B}^{-1}$ we let $a \in \mathbb{R}$ and let $f \in H_{\varrho,0}(\mathbb{R}; H)$ with $\chi_{]-\infty,a]} f = 0$. We have to prove that $\chi_{]-\infty,a]} \overline{B}^{-1} f = 0$. We put $U := \overline{B}^{-1} f$ and use (1.3.9) with $\overline{B}$ instead of B to obtain

$$0 = \langle \chi_{]-\infty,a]} f | U \rangle_{\varrho,0} = \langle f | \chi_{]-\infty,a]} U \rangle_{\varrho,0}$$
$$= \langle \overline{B} U | \chi_{]-\infty,a]} U \rangle_{\varrho,0} \geq c_0 \langle \chi_{]-\infty,a]} U | \chi_{]-\infty,a]} U \rangle_{\varrho,0}.$$

Thus, $\chi_{]-\infty,a]} U = \chi_{]-\infty,a]} \overline{B}^{-1} f = 0$, which yields the assertion. $\qquad \square$

Theorem 1.3.6 (Non-dependence on ρ) *Under the assumptions of Theorem 1.3.2, let $\varrho_1, \varrho_2 \in [\varrho_0, \infty[$ and $f \in H_{\varrho_1,0}(\mathbb{R}, H) \cap H_{\varrho_2,0}(\mathbb{R}, H)$ with $\varrho_1 \leq \varrho_2$. For $k \in \{1, 2\}$, denote by $U_{\varrho_k} \in H_{\varrho_k,0}(\mathbb{R}; H)$ the solution of*

$$\big(\partial_{0,\varrho_k} M_0 + M_1 + A\big) U_{\varrho_k} = f.$$

Then $U_{\varrho_1} = U_{\varrho_2}$.

For the proof, we need the following auxiliary result.

Lemma 1.3.7 *Let $u \in H_{\varrho_1,0}(\mathbb{R}; H) \cap H_{\varrho_2,0}(\mathbb{R}; H)$ for some $\varrho_1, \varrho_2 \in \mathbb{R}$.*

(a) If $u \in \mathrm{dom}(\partial_{0,\varrho_2})$, then $u \in \mathrm{dom}(\partial_{0,\varrho_1})$ with $\partial_{0,\varrho_1} u = \partial_{0,\varrho_2} u$.

(b) If $\varrho_1, \varrho_2 > 0$, then

$$(1 + \varepsilon \partial_{0,\varrho_1})^{-1} u = (1 + \varepsilon \partial_{0,\varrho_2})^{-1} u \quad (\varepsilon > 0).$$

Proof

(a) We prove that $u \in \mathrm{dom}(\partial_{0,\varrho_1}) = \mathrm{dom}(\partial_{0,\varrho_1}^*)$. For $\varphi \in \overset{\circ}{C}_1(\mathbb{R}; H)$ we compute

$$\langle u | \partial_{0,\varrho_1} \varphi \rangle_{\varrho_1,0}$$
$$= \langle u | \varphi' \rangle_{\varrho_1,0}$$
$$= \langle u | \exp(2(\varrho_2 - \varrho_1)\cdot)\varphi' \rangle_{\varrho_2,0}$$
$$= \langle u | (\exp(2(\varrho_2 - \varrho_1)\cdot)\varphi)' - 2(\varrho_2 - \varrho_1) \exp(2(\varrho_2 - \varrho_1)\cdot)\varphi \rangle_{\varrho_2,0}$$
$$= \langle \partial_{0,\varrho_2}^* u - 2(\varrho_2 - \varrho_1) u | \varphi \rangle_{\varrho_1,0}$$
$$= \langle -\partial_{0,\varrho_2} u + 2\varrho_1 u | \varphi \rangle_{\varrho_1,0},$$

where we have used Corollary 1.1.5. Thus $u \in \mathrm{dom}(\partial_{0,\varrho_1}^*)$ with

$$\partial_{0,\varrho_1}^* u = -\partial_{0,\varrho_2} u + 2\varrho_1 u.$$

Again, Corollary 1.1.5 yields $\partial_{0,\varrho_1} u = \partial_{0,\varrho_2} u$.

(b) Without loss of generality, we assume that $\varrho_1 \leq \varrho_2$. Let $n \in \mathbb{N}$ and set $u_n := \chi_{[-n,\infty[} u$. Then by dominated convergence, $u_n \to u$ in $H_{\varrho_1,0}(\mathbb{R}; H)$ and $H_{\varrho_2,0}(\mathbb{R}; H)$. Moreover, we note that $(1 + \varepsilon \partial_{0,\varrho_1})^{-1}$ is causal by Theorem 1.3.2 (with $M_0 = \varepsilon$, $M_1 = 1$ and $A = 0$) and thus,

$$f_n := (1 + \varepsilon \partial_{0,\varrho_1})^{-1} u_n \in H_{\varrho_1,0}(\mathbb{R}; H)$$

satisfies $\mathrm{supp}\, f_n \subseteq [-n, \infty[$ and hence, $f_n \in H_{\varrho_2,0}(\mathbb{R}; H)$. Consequently,

$$\partial_{0,\varrho_1} f_n = \frac{1}{\varepsilon}(u_n - f_n) \in H_{\varrho_1,0}(\mathbb{R}; H)$$

and thus, $f_n \in \mathrm{dom}(\partial_{0,\varrho_2})$ with

$$\partial_{0,\varrho_2} f_n = \frac{1}{\varepsilon}(u_n - f_n)$$

according to (a). Summarizing, we have shown that

$$(1 + \varepsilon \partial_{0,\varrho_1})^{-1} u_n = f_n = (1 + \varepsilon \partial_{0,\varrho_2})^{-1} u_n$$

for each $n \in \mathbb{N}$. Taking the limit as $n \to \infty$, we derive the assertion. $\square$

Now we can address the proof of Theorem 1.3.6.

Proof of Theorem 1.3.6 Let $a \in \mathbb{R}_{>0}$ and $\varphi \in C_\infty (\mathbb{R})$ be such that $\varphi = 1$ on $]-\infty, a[$ and $\varphi = 0$ on $]a + 1, \infty[$. Given $\varepsilon > 0$, define $U_{\varrho k,\varepsilon} := (1 + \varepsilon\partial_{0,\varrho k})^{-1} U_{\varrho k}$ for $k \in \{1, 2\}$. Then

$$\left(\partial_{0,\varrho_1} M_0 + M_1 + A\right) \varphi U_{\varrho 1,\varepsilon} = \varphi(1 + \varepsilon\partial_{0,\varrho_1})^{-1} f + \varphi' M_0 U_{\varrho 1,\varepsilon}$$

$$\left(\partial_{0,\varrho_2} M_0 + M_1 + A\right) \varphi U_{\varrho 2,\varepsilon} = \varphi(1 + \varepsilon\partial_{0,\varrho_2})^{-1} f + \varphi' M_0 U_{\varrho 2,\varepsilon}$$

in $H_{\varrho_1} (\mathbb{R}; H)$ and $H_{\varrho_2} (\mathbb{R}; H)$, respectively. We shall show that

$$\varphi U_{\varrho 2,\varepsilon} \in H_{\varrho_1,0} (\mathbb{R}; H) .$$

We estimate—similar to Lemma 1.2.6—

$$\left|\varphi U_{\varrho 2,\varepsilon}\right|^2_{\varrho_1,0} = \int_{-\infty}^{\infty} \left|\varphi (t) U_{\varrho 2,\varepsilon} (t)\right|^2_0 \exp(-2\varrho_1 t)\, dt$$

$$= \int_{-\infty}^{a+1} \left|\varphi (t) U_{\varrho 2,\varepsilon} (t)\right|^2_0 \exp(-2\varrho_1 t)\, dt$$

$$= \exp(-2\varrho_1 (a + 1)) \int_{-\infty}^{a+1} \left|\varphi (t) U_{\varrho 2,\varepsilon} (t)\right|^2_0 \exp(-2\varrho_1 (t - a - 1))\, dt$$

$$\leq \exp(-2\varrho_1 (a + 1)) \int_{-\infty}^{a+1} \left|\varphi (t) U_{\varrho 2,\varepsilon} (t)\right|^2_0 \exp(-2\varrho_2 (t - a - 1))\, dt$$

$$\leq \exp(2 (\varrho_2 - \varrho_1) (a + 1)) \int_{-\infty}^{a+1} \left|\varphi (t) U_{\varrho 2,\varepsilon} (t)\right|^2_0 \exp(-2\varrho_2 t)\, dt$$

$$\leq \exp(2 (\varrho_2 - \varrho_1) (a + 1)) \left|\varphi U_{\varrho 2,\varepsilon}\right|^2_{\varrho_2,0} .$$

In the same way we get $\varphi' U_{\varrho 2,\varepsilon} \in H_{\varrho_1,0}(\mathbb{R}; H)$ and hence

$$\partial_{0,\varrho_2} M_0 U_{\varrho 2,\varepsilon} = \varphi(1 + \varepsilon\partial_{0,\varrho_2})^{-1} f + \varphi' M_0 U_{\varrho 2,\varepsilon} - M_1 \varphi U_{\varrho 2,\varepsilon} - A\varphi U_{\varrho 2,\varepsilon} \in H_{\varrho_1,0}(\mathbb{R}; H).$$

Thus, employing Lemma 1.3.7, we have in $H_{\varrho_1,0} (\mathbb{R}, H)$

$$\left(\partial_{0,\varrho_1} M_0 + M_1 + A\right) \varphi \left(U_{\varrho 2,\varepsilon} - U_{\varrho 1,\varepsilon}\right) = \varphi' M_0 \left(U_{\varrho 2,\varepsilon} - U_{\varrho 1,\varepsilon}\right) .$$

Since the right-hand side vanishes on $]-\infty, a[$ we have by causality that

$$U_{\varrho 1,\varepsilon} = U_{\varrho 2,\varepsilon} \text{ on }]-\infty, a[.$$

Since $a \in {]}0, \infty{[}$ is arbitrary, we obtain $U_{\varrho_1, \varepsilon} = U_{\varrho_2, \varepsilon}$ for all $\varepsilon > 0$. Hence, by Corollary 1.1.8 and the Fischer–Riesz theorem, we obtain for almost all $t \in \mathbb{R}$ and a null-sequence $(\varepsilon_j)_j$ in $\mathbb{R}_{>0}$:

$$U_{\varrho_1}(t) = \lim_{j \to \infty} U_{\varrho_1, \varepsilon_j}(t) = \lim_{j \to \infty} U_{\varrho_2, \varepsilon_j}(t) = U_{\varrho_2}(t),$$

which yields the assertion.   □

Before we elaborate on a more involved perturbation theorem, we will state a first straightforward consequence of the results obtained so far.

Theorem 1.3.8 *Let $f \colon \mathrm{dom}(f) \subseteq \bigcap_{\varrho \geq \varrho_0} H_{\varrho,0}(\mathbb{R}; H) \to \bigcap_{\varrho \geq \varrho_0} H_{\varrho,0}(\mathbb{R}; H)$ be evolutionary at $\varrho_0 \geq 0$ and causal. Then, for all sufficiently large ϱ, and for all $F \in H_{\varrho,0}(\mathbb{R}; H)$ the equation*

$$\overline{(\partial_0 M_0 + M_1 + A)}U = \partial_0^{-1} f_\varrho(U) + F$$

admits a unique solution $U \in H_{\varrho,0}(\mathbb{R}; H)$, the solution depends Lipschitz continuously on F and the mapping $F \mapsto U$ is causal.

Proof Note that by Theorem 1.3.2, the operator $\overline{(\partial_0 M_0 + M_1 + A)}$ is continuously invertible for all ϱ sufficiently large and the norm of the inverse is uniformly bounded for all sufficiently large ϱ. Next, as f is evolutionary and $\|\partial_{0,\varrho}^{-1}\| \leq 1/\varrho$, we obtain that $\left|\partial_{0,\varrho}^{-1} f_\varrho\right|_{\mathrm{Lip}} \to 0$ as $\varrho \to \infty$. Thus, the assertion eventually follows from the contraction mapping principle. (Note that the composition of causal mappings is again causal.)   □

1.3.3 Lipschitz Continuous Perturbations

So far we have avoided more complicated material laws to keep matters elementary. The following perturbation result, however, allows the inclusion of a large class of more involved material laws; a refinement of this approach leads to well-posedness results for stochastic partial differential equations, see [72, 76].

We first need to sharpen our positivity assumption for the material law operator in (1.3.1). Similar to Sect. 1.3.1, we assume that $M_0 = M_0^*$, $M_1 \in \mathcal{B}(H)$ and $A = -A^*$ in H as well as the following conditions:

- M_0 is strictly positive on its range, that is, there exists $c_1 \in \mathbb{R}_{>0}$ such that for all $\varphi \in M_0[H] = \mathrm{ran}(M_0)$

$$\langle M_0 \varphi | \varphi \rangle_H \geq c_1 \langle \varphi | \varphi \rangle_H \tag{1.3.10}$$

- M_1 is strictly positive on the null space of M_0, that is, there exists $c_2 \in \mathbb{R}_{>0}$ such that for all $\psi \in [\{0\}]M_0 = \ker(M_0)$

$$\langle M_1 \psi | \psi \rangle_H \geq c_2 \langle \psi | \psi \rangle_H. \tag{1.3.11}$$

Next, we confirm that the last two assumptions indeed imply (1.3.1) for suitable $\varrho_0, c_0 > 0$. For this we denote

$$\iota_r \colon M_0[H] \hookrightarrow H$$

$$\iota_n \colon [\{0\}]M_0 \hookrightarrow H,$$

the canonical embeddings, and realize that

$$\iota_r^* \colon H \to M_0[H]$$

$$\iota_n^* \colon H \to [\{0\}]M_0$$

act as orthogonal projections corresponding to the orthogonal decomposition $H = M_0[H] \oplus [\{0\}]M_0$, which is implied by the selfadjointness of M_0 and the closedness of $M_0[H]$, which follows from the positivity estimate imposed on M_0, see also Lemma B.4.18 or [66, Lemma 3.2]. We set

$$P_r := \iota_r \iota_r^* \text{ and } P_n := \iota_n \iota_n^*,$$

which are the selfadjoint orthogonal projections on $M_0[H]$ and $[\{0\}]M_0$, respectively. The above positivity estimates imposed on M_0 and M_1 imply a more refined estimate than the one in (1.3.1):

Proposition 1.3.9 *Let $M_0 = M_0^*$, $M_1 \in \mathcal{B}(H)$ satisfy (1.3.10) and (1.3.11). Then for all $\varrho \in \mathbb{R}_{>0}$ and $\varepsilon \in {]0, 1[}$ we have, for all $U \in H$,*

$$\langle U | (\varrho M_0 + M_1) U \rangle_H \geq \kappa (\varrho, \varepsilon) | P_r U |_H^2 + c_2 (1 - \varepsilon) | P_n U |_H^2,$$

where $\kappa(\varrho, \varepsilon) := \varrho c_1 - \| P_r M_1 P_r \| - (1/(4\varepsilon c_2)) (\| P_r M_1 P_n \| + \| P_n M_1 P_r \|)^2.$

Proof Let $\varrho \in \mathbb{R}_{>0}$, $\varepsilon \in {]0, 1[}$. Then we estimate with the help of (1.3.10) and (1.3.11) as well as Euclid's inequality $(-2ab \geq -a^2 - b^2, a, b \in \mathbb{R})$:

$$\langle U | (\varrho M_0 + M_1) U \rangle_H$$

$$= \varrho \langle P_r U | M_0 P_r U \rangle_H$$

$$+ \langle P_r U | M_1 P_r U \rangle_H + \langle P_r U | M_1 P_n U \rangle_H + \langle P_n U | M_1 P_r U \rangle_H + \langle P_n U | M_1 P_n U \rangle_H$$

$$\geq (\varrho c_1 - \|P_r M_1 P_r\|)\, |P_r U|_H^2$$

$$- (\|P_r M_1 P_n\| + \|P_n M_1 P_r\|)\, |P_r U|_H |P_n U|_H + c_2 |P_n U|_H^2$$

$$\geq \left(\varrho c_1 - \|P_r M_1 P_r\| - \frac{1}{4\varepsilon c_2}(\|P_r M_1 P_n\| + \|P_n M_1 P_r\|)^2\right) |P_r U|_H^2$$

$$+ c_2(1 - \varepsilon)|P_n U|_H^2. \qquad \qquad \square$$

This result yields (1.3.1) for

$$c_0 := \min\{\kappa(\varrho, \varepsilon), c_2(1 - \varepsilon)\}$$

as long as ϱ is sufficiently large, that is, so large such that $\kappa(\varrho, \varepsilon) > 0$ or, equivalently,

$$\varrho > \frac{\|P_r M_1 P_r\|}{c_1} + \frac{(\|P_r M_1 P_n\| + \|P_n M_1 P_r\|)^2}{4\varepsilon c_1 c_2}. \qquad (1.3.12)$$

Proposition 1.3.9 yields a refinement of the continuity estimate of the solution operator:

Proposition 1.3.10 *Let $M_0 = M_0^*$, $M_1 \in L(H)$ satisfy (1.3.10) and (1.3.11). Let $A = -A^*$ in H, $\varepsilon \in \,]0, 1[$. Then, for all*

$$\varrho > \varrho_0 := \frac{\|P_r M_1 P_r\|}{c_1} + \frac{(\|P_r M_1 P_n\| + \|P_n M_1 P_r\|)^2}{4\varepsilon c_1 c_2} + \frac{c_2}{c_1}\frac{1 - \varepsilon}{2}$$

and $U \in H_{\varrho,0}(\mathbb{R}; H)$ such that

$$\overline{(\partial_0 M_0 + M_1 + A)}U = F \in H_{\varrho,0}(\mathbb{R}; H)$$

we have

$$|F|_{\varrho,0}^2 \geq c_2(1 - \varepsilon)(2\kappa(\varrho, \varepsilon) - c_2(1 - \varepsilon))\,|P_r U|_{\varrho,0}^2 + (c_2(1 - \varepsilon))^2\,|P_n U|_{\varrho,0}^2,$$

where $\kappa(\varrho, \varepsilon)$ is given in Proposition 1.3.9.

Proof We put $\varepsilon_1 := 1 - \varepsilon$ and estimate with (1.3.5) and Euclid's inequality:

$$\frac{1}{2}\left(c_2\varepsilon_1|U|_{\varrho,0}^2 + \frac{1}{c_2\varepsilon_1}|F|_{\varrho,0}^2\right) \geq \langle U|F\rangle_{\varrho,0}$$

$$= \langle U|\,\overline{(\partial_0 M_0 + M_1 + A)}\,U\rangle_{\varrho,0} \geq \langle U|\,(\varrho M_0 + M_1)\,U\rangle_{\varrho,0}.$$

We estimate further with the help of Proposition 1.3.9 and obtain

$$\frac{1}{2}\left(c_2\varepsilon_1|U|^2_{\varrho,0} + \frac{1}{c_2\varepsilon_1}|F|^2_{\varrho,0}\right) \geq \langle U\,|\,(\varrho M_0 + M_1)\,U\rangle_{\varrho,0}$$

$$\geq \kappa\,(\varrho,\varepsilon)\,|P_{\mathrm{r}}U|^2_{\varrho,0} + c_2\,(1-\varepsilon)\,|P_{\mathrm{n}}U|^2_{\varrho,0}.$$

Hence, we get

$$\frac{1}{2c_2\varepsilon_1}\,|F|^2_{\varrho,0} \geq \left(\kappa\,(\varrho,\varepsilon) - c_2\frac{1-\varepsilon}{2}\right)|P_{\mathrm{r}}U|^2_{\varrho,0} + c_2\left(\frac{1-\varepsilon}{2}\right)|P_{\mathrm{n}}U|^2_{\varrho,0},$$

which is equivalent to the assertion. $\square$

With these preliminaries at hand, we are now able to show the following perturbation result.

Theorem 1.3.11 *Let $M_0 = M_0^*$, $M_1 \in \mathcal{B}(H)$ satisfy (1.3.10) and (1.3.11). Assume A to be skew-selfadjoint in H, $\varrho_1 \in \mathbb{R}$. Moreover, let M_2 be evolutionary at ϱ_1 and satisfy the following refined Lipschitz condition: There exist $L_0, L_1 \in \,]0, \infty[$ with $L_1 < c_2$ such that for all $V_0, V_1 \in \mathrm{dom}(M_2)$ and $\varrho \in \mathbb{R}_{>\varrho_1}$:*

$$|M_2\,(V_0) - M_2\,(V_1)|^2_{\varrho,0} \leq L_0^2\,|P_{\mathrm{r}}V_0 - P_{\mathrm{r}}V_1|^2_{\varrho,0} + L_1^2\,|P_{\mathrm{n}}V_0 - P_{\mathrm{n}}V_1|^2_{\varrho,0}$$

Then there exists $\varrho_0 \in \mathbb{R}_{>\varrho_1}$ such that for all $\varrho \in \mathbb{R}_{\geq\varrho_0}$ the equation

$$\left(\partial_{0,\varrho}M_0 + M_1 + M_2\,(\,\cdot\,) + A\right)U = F \in H_{\varrho,0}\,(\mathbb{R}, H)$$

admits a unique solution $U \in H_{\varrho,0}(\mathbb{R}; H)$.

Moreover, we have continuous dependence on the data in the sense that the Lipschitz semi-norm of the solution operator is bounded, that is,

$$\left|\left(\partial_{0,\varrho}M_0 + M_1 + M_{2,\varrho}\,(\,\cdot\,) + A\right)^{-1}\right|_{\mathrm{Lip}} < \infty.$$

Furthermore, if M_2 is causal, then

$$\left(\partial_{0,\varrho}M_0 + M_1 + M_{2,\varrho}\,(\,\cdot\,) + A\right)^{-1} =$$

$$\left(1 + \left(\partial_{0,\varrho}M_0 + M_1 + A\right)^{-1}M_{2,\varrho}\,(\,\cdot\,)\right)^{-1}\left(\partial_{0,\varrho}M_0 + M_1 + A\right)^{-1} \qquad (1.3.13)$$

is also causal and the solution operator is independent of ϱ.

Proof Let $\varepsilon \in \,]0, 1[$ be such that

$$L_1 < c_2(1 - \varepsilon) \tag{1.3.14}$$

and let $\widetilde{\varrho_0}$ be defined as ϱ_0 from Proposition 1.3.10. Then, all $\varrho > \widetilde{\varrho_0}$ satisfy (1.3.12) and hence, by Theorem 1.3.2, the mapping $\left(\partial_{0,\varrho} M_0 + M_1 + A\right)^{-1} \in L(H_{\varrho,0}(\mathbb{R}; H))$ is well-defined. Recall $\kappa(\varrho, \varepsilon)$ as given in Proposition 1.3.9 and choose $\varrho_0 \in \mathbb{R}_{>\widetilde{\varrho_0}}$ with $\varrho_0 > \varrho_1$ such that

$$\frac{L_0^2}{c_2 \, (1 - \varepsilon) \, (2\kappa \, (\varrho_0, \varepsilon) - c_2 \, (1 - \varepsilon))} < 1. \tag{1.3.15}$$

Let $\rho \geq \rho_0$. For $F \in H_{\varrho,0}(\mathbb{R}; H)$ we define

$$\Phi_{\varrho,F} : H_{\varrho,0}(\mathbb{R}; H) \to H_{\varrho,0}(\mathbb{R}; H)$$

$$U \mapsto \left(\partial_{0,\varrho} M_0 + M_1 + A\right)^{-1} \left(F - M_{2,\varrho}(U)\right),$$

where $M_{2,\varrho}$ is the closure of M_2 in $H_{\varrho,0}(\mathbb{R}; H)$. We note that for all $U \in H_{\varrho,0}(\mathbb{R}; H)$

$$\Phi_{\varrho,F}(U) = U \iff \left(\partial_{0,\varrho} M_0 + M_1 + A\right) U = F - M_{2,\varrho}(U)$$

$$\iff \left(\partial_{0,\varrho} M_0 + M_1 + M_{2,\varrho}(\cdot) + A\right) U = F,$$

which also implies (1.3.13). Thus, we are left with showing that $\Phi_{\varrho,F}$ admits a unique fixed point in $H_{\varrho,0}(\mathbb{R}; H)$. For this let $V_0, V_1 \in H_{\varrho,0}(\mathbb{R}; H)$ and define $U_k := \Phi_{\varrho,F}(V_k)$, $k \in \{0, 1\}$ and $U := U_0 - U_1$. Then

$$\left(\partial_{0,\varrho} M_0 + M_1 + A\right) U = M_{2,\varrho}(V_1) - M_{2,\varrho}(V_0)$$

and Proposition 1.3.10 and the hypothesis on M_2, we obtain

$$L_0^2 \, |P_\mathrm{r} V_0 - P_\mathrm{r} V_1|_{\varrho,0}^2 + L_1^2 \, |P_\mathrm{n} V_0 - P_\mathrm{n} V_1|_{\varrho,0}^2$$

$$\geq \left|M_{2,\varrho} \, (V_0) - M_{2,\varrho} \, (V_1)\right|_{\varrho,0}^2$$

$$\geq c_2 \, (1 - \varepsilon) \, (2\kappa \, (\varrho, \varepsilon) - c_2 \, (1 - \varepsilon)) \, |P_\mathrm{r} U|_{\varrho,0}^2 + (c_2 \, (1 - \varepsilon))^2 \, |P_\mathrm{n} U|_{\varrho,0}^2 \, .$$

Defining

$$\lambda_0 := c_2 \, (1 - \varepsilon) \, (2\kappa \, (\varrho, \varepsilon) - c_2 \, (1 - \varepsilon)) \, ,$$

$$\lambda_1 := (c_2 \, (1 - \varepsilon))^2 \, , \text{ and}$$

$$\kappa_* := \max\left\{\frac{L_0^2}{\lambda_0}, \frac{L_1^2}{\lambda_1}\right\},$$

we see by (1.3.14) and (1.3.15) that $\kappa_* < 1$. Moreover, we obtain

$$\kappa_*\left(\lambda_0 |P_r V_0 - P_r V_1|_{\varrho,0}^2 + \lambda_1 |P_n V_0 - P_n V_1|_{\varrho,0}^2\right)$$

$$\geq \lambda_0 |P_r U|_{\varrho,0}^2 + \lambda_1 |P_n U|_{\varrho,0}^2.$$

$$= \lambda_0 |P_r U_0 - P_r U_1|_{\varrho,0}^2 + \lambda_1 |P_n U_0 - P_n U_1|_{\varrho,0}^2$$

Thus, $\Phi_{\varrho,F}$ is a contraction in $H_{\varrho,0}(\mathbb{R}, H)$ with respect to

$$|U|_\lambda := \sqrt{\lambda_0 |P_r U|_{\varrho,0}^2 + \lambda_1 |P_n U|_{\varrho,0}^2}$$

as (equivalent) norm. Hence, unique existence of a fixed point U of $\Phi_{\varrho,F}$ follows from Banach's fixed point theorem.[5]

Next, we address the continuous dependence on the data. For this, let $F_0, F_1 \in H_{\varrho,0}(\mathbb{R}; H)$ and let U_0 and U_1 be the fixed points of Φ_{ϱ,F_0} and Φ_{ϱ,F_1}. We estimate

$$|U_0 - U_1|_\lambda$$

$$= \left|\Phi_{\varrho,F_0}(U_0) - \Phi_{\varrho,F_1}(U_1)\right|_\lambda$$

$$= \left|\left(\partial_{0,\varrho} M_0 + M_1 + A\right)^{-1}\left(F_0 - M_{2,\varrho}(U_0)\right) - \left(\partial_{0,\varrho} M_0 + M_1 + A\right)^{-1}\left(F_1 - M_{2,\varrho}(U_1)\right)\right|_\lambda$$

$$\leq \left\|\left(\partial_{0,\varrho} M_0 + M_1 + A\right)^{-1}\right\| |F_0 - F_1|_\lambda + \left|\Phi_{\varrho,0}(U_0) - \Phi_{\varrho,0}(U_1)\right|_\lambda$$

$$\leq \left\|\left(\partial_{0,\varrho} M_0 + M_1 + A\right)^{-1}\right\| |F_0 - F_1|_\lambda + |\Phi_{\varrho,0}|_{\text{Lip}} |U_0 - U_1|_\lambda$$

from which we obtain

$$|U_0 - U_1|_\lambda \leq \frac{\left\|\overline{\partial_{0,\varrho} M_0 + M_1 + A}^{-1}\right\|}{1 - |\Phi_{\varrho,0}|_{\text{Lip}}} |F_0 - F_1|_\lambda$$

---

[5] As a by-product we recall that for arbitrary choice of $V \in H_{\varrho,0}(\mathbb{R}, H)$ we have

$$\left|\Phi_{\varrho,0}^k(V) - U\right|_\lambda \leq \frac{|\Phi_{\varrho,0}|_{\text{Lip}}^k}{1 - |\Phi_{\varrho,0}|_{\text{Lip}}} |\Phi_{\varrho,0}(V) - U|_\lambda.$$

and so

$$\left| \overline{\partial_{0,\varrho} M_0 + M_1 + M_2\,(\;\cdot\;) + A}^{\,-1} \right|_{\mathrm{Lip}} \leq \frac{\left\| \overline{\partial_{0,\varrho} M_0 + M_1 + A}^{\,-1} \right\|}{1 - \left| \Phi_{\varrho,0} \right|_{\mathrm{Lip}}}.$$

Causality follows for a causal M_2 by the causality of the fix point mapping. Indeed, let $F, G \in H_{\varrho,0}(\mathbb{R}; H)$ with $F = G$ on $]-\infty, a]$ for some $a \in \mathbb{R}$. Then, from the causality of M_2 and of $\left(\partial_{0,\varrho} M_0 + M_1 + A \right)^{-1}$ we get

$$\chi_{]-\infty,a]} \Phi_{\varrho,F}(U) = \chi_{]-\infty,a]} \Phi_{\varrho,F}(\chi_{]-\infty,a]}U) = \chi_{]-\infty,a]} \Phi_{\varrho,G}(\chi_{]-\infty,a]}U)$$

for each $U \in H_{\varrho,0}(\mathbb{R}; H)$ and thus, denoting the fixed point of $\Phi_{\varrho,F}$ by U_F, we get

$$\chi_{]-\infty,a]} U_F = \chi_{]-\infty,a]} \Phi_{\varrho,F}(U_F) = \chi_{]-\infty,a]} \Phi_{\varrho,G}(\chi_{]-\infty,a]}U_F).$$

Hence, due to the uniqueness of the fixed point of $\chi_{]-\infty,a]} \Phi_{\varrho,G}$ we get

$$\chi_{]-\infty,a]} U_F = \chi_{]-\infty,a]} U_G.$$

It is left to prove the non-dependence on the parameter ϱ. Let $F \in H_{\varrho,0}(\mathbb{R}; H) \cap H_{\mu,0}(\mathbb{R}; H)$ for some $\varrho, \mu \geq \varrho_0$ and denote by $U_\varrho \in H_{\varrho,0}(\mathbb{R}; H)$ and $U_\mu \in H_{\mu,0}(\mathbb{R}; H)$ the respective solutions. Then we have (up to a subsequence)

$$U_\varrho(t) = \lim_{n \to \infty} \left(\Phi_{\varrho,F}^n(0) \right)(t)$$

$$U_\mu(t) = \lim_{n \to \infty} \left(\Phi_{\mu,F}^n(0) \right)(t)$$

for almost all $t \in \mathbb{R}$. Now, due to the independence statement in Theorem 1.3.6 and the non-dependence on ϱ of $M_{2,\varrho}$ by Lemma 1.2.6, we get

$$\Phi_{\varrho,F}(V) = \left(\partial_{0,\varrho} M_0 + M_1 + A \right)^{-1} \left(F - M_{2,\varrho}(V) \right)$$

$$= \left(\partial_{0,\mu} M_0 + M_1 + A \right)^{-1} \left(F - M_{2,\mu}(V) \right)$$

$$= \Phi_{\mu,F}(V)$$

for each $V \in H_{\varrho,0}(\mathbb{R}; H) \cap H_{\mu,0}(\mathbb{R}; H)$ and hence, by induction

$$\Phi_{\varrho,F}^n(0) = \Phi_{\mu,F}^n(0)$$

for each $n \in \mathbb{N}$, which yields the assertion. $\square$

Some Applications to Models from Physics and Engineering

2

A vast number of physical and engineering models can be shown to give rise to equations involving an operator of the form

$$\partial_0 M_0 + M_1 + A \tag{2.0.1}$$

where A is a skew-selfadjoint operator of the form

$$A = \begin{pmatrix} 0 & -C^* \\ C & 0 \end{pmatrix}, \tag{2.0.2}$$

with $C : \operatorname{dom}(C) \subseteq H_0 \to H_1$ being a densely defined, closed, linear operator from a Hilbert space H_0 into a possibly different Hilbert space H_1. It is straightforward to show that A is skew-selfadjoint in $H := H_0 \oplus H_1$ (see Corollary B.4.16). In (2.0.1), the operators M_0 and M_1 are bounded linear operators in H. In order to maintain continuous invertibility of (the closure of) $\partial_0 M_0 + M_1 + A$ in $H_{\varrho,0}(\mathbb{R}; H)$ for some $\varrho > 0$ in applications, we will often be confronted with showing that M_0 and M_1 satisfy the conditions mentioned in (1.3.10) and (1.3.11).

In connection with basic classical phenomena of mathematical physics and engineering we have as particular choices for C operators associated with the classical vector analysis, operations grad, div, curl as well as their analogues for higher tensors. In the light of this, there is in applications only a limited number of different operators A. Plethora of equations comes into play through the multitude of different constitutive relations, which find their expression in various choices for the operators M_0 and M_1, which we shall call *material law operators* or just *material laws*.

However, from a mathematical point of view there are fewer equations than one would think judging from the huge variety of phenomena being considered in various fields of

© Springer Nature Switzerland AG 2020
R. Picard et al., *A Primer for a Secret Shortcut to PDEs of Mathematical Physics*,
Frontiers in Mathematics, https://doi.org/10.1007/978-3-030-47333-4_2

applications. It turns out that frequently mathematically equivalent models are appearing under different names (possibly with different units and different interpretations). We shall comment on this in passing in the following discussion of some typical cases. Many more examples, as well as some further analysis (such as a suitable framework for exponential stability or homogenization) can be found in [68]; see also [81, 88] and the references therein.

2.1   Acoustic Equations and Related Problems

We begin with the case of operators $\overset{\circ}{\mathrm{grad}}$ and $\overset{\circ}{\mathrm{div}}$ obtained from the classical vector analytical operations by taking the L^2-closure of the gradient (grad) of smooth functions and of the divergence (div) of smooth vector fields with compact support, that is, vanishing outside a compact set in the underlying domain Ω to which the physical processes under consideration are thought to be confined. More formally we have:

Definition 2.1.1 Let $\Omega \subseteq \mathbb{R}^3$ be open.[1] We define

$$\mathrm{grad}\,|_{\overset{\circ}{C}_\infty(\Omega)} : \overset{\circ}{C}_\infty(\Omega) \subseteq L^2(\Omega) \to L^2(\Omega)^3$$

$$\varphi \mapsto \mathrm{grad}\,\varphi := \begin{pmatrix} \partial_1\varphi \\ \partial_2\varphi \\ \partial_3\varphi \end{pmatrix}$$

and put

$$\overset{\circ}{\mathrm{grad}} := \overline{\mathrm{grad}\,|_{\overset{\circ}{C}_\infty(\Omega)}}.$$

Similarly, we set

$$\mathrm{div}\,|_{\overset{\circ}{C}_\infty(\Omega)^3} : \overset{\circ}{C}_\infty(\Omega)^3 \subseteq L^2(\Omega)^3 \to L^2(\Omega)$$

$$\begin{pmatrix} \Psi_1 \\ \Psi_2 \\ \Psi_3 \end{pmatrix} \mapsto \mathrm{div}\begin{pmatrix} \Psi_1 \\ \Psi_2 \\ \Psi_3 \end{pmatrix} := \partial_1\Psi_1 + \partial_2\Psi_2 + \partial_3\Psi_3,$$

and define

$$\overset{\circ}{\mathrm{div}} := \overline{\mathrm{div}\,|_{\overset{\circ}{C}_\infty(\Omega)^3}}.$$

---

[1]For sake of simplicity, we restrict our attention to three space dimensions. The higher dimensional case can be dealt with in an analogous fashion.

Remark 2.1.2 We note here that these definitions of $\overset{\circ}{\operatorname{div}}$ and $\overset{\circ}{\operatorname{grad}}$ remain the same if $\overset{\circ}{C}_\infty(\Omega)$ were to be replaced by $\overset{\circ}{C}_1(\Omega)$. Moreover, it is so far not clear that $\overset{\circ}{\operatorname{grad}}$ and $\overset{\circ}{\operatorname{div}}$ are indeed operators. However, the next proposition will show that this is indeed the case.

Proposition 2.1.3 *Let $\Omega \subseteq \mathbb{R}^3$ be open. Then $\overset{\circ}{\operatorname{grad}}$ and $\overset{\circ}{\operatorname{div}}$ are formal skew-adjoints of each other, that is,*

$$\overset{\circ}{\operatorname{grad}} \subseteq -\operatorname{div}^*$$

or, equivalently,

$$\overset{\circ}{\operatorname{div}} \subseteq -\operatorname{grad}^*.$$

In particular, $\overset{\circ}{\operatorname{grad}}$ and $\overset{\circ}{\operatorname{div}}$ are densely defined closed linear operators.

Proof Let $\varphi \in \overset{\circ}{C}_\infty(\Omega) = \operatorname{dom}(\operatorname{grad}|_{\overset{\circ}{C}_\infty(\Omega)})$ and $\Psi \in \overset{\circ}{C}_\infty(\Omega)^3$. We extend both φ and Ψ by zero on $\mathbb{R}^3 \setminus \Omega$. Note that then $\varphi \in \overset{\circ}{C}_\infty(\mathbb{R}^3)$ and $\Psi \in \overset{\circ}{C}_\infty(\mathbb{R}^3)^3$. Then there exists $R \in \,]0, \infty[$ such that $\operatorname{supp}\varphi$, $\operatorname{supp}\Psi \subseteq B(0, R)$. By integration by parts (Gauss' theorem), we find

$$\left\langle \overset{\circ}{\operatorname{grad}}\varphi \,\middle|\, \Psi \right\rangle_{L^2(\Omega)^3} + \left\langle \varphi \,\middle|\, \overset{\circ}{\operatorname{div}}\Psi \right\rangle_{L^2(\Omega)} \tag{2.1.1}$$

$$= \langle \operatorname{grad}\varphi | \Psi \rangle_{L^2(B(0,R))^3} + \langle \varphi | \operatorname{div}\Psi \rangle_{L^2(B(0,R))}$$

$$= \int_{B(0,R)} (\partial_1\varphi(x)\Psi_1(x) + \partial_2\varphi(x)\Psi_2(x) + \partial_3\varphi(x)\Psi_3(x))\,dx +$$

$$+ \int_{B(0,R)} (\varphi(x)\partial_1\Psi_1(x) + \varphi(x)\partial_2\Psi_2(x) + \varphi(x)\partial_3\Psi_3(x))\,dx$$

$$= \int_{B(0,R)} \operatorname{div}(\varphi\Psi)\,dV$$

$$= \int_{\partial B(0,R)} \left(\varphi(x)\frac{x_1}{R}\Psi_1(x) + \varphi(x)\frac{x_2}{R}\Psi_2(x) + \varphi(x)\frac{x_3}{R}\Psi_3(x)\right)dS(x)$$

$$= 0$$

where dV is the volume element, here of $B(0, R)$, and dS denotes the surface element, here of the sphere $\partial B(0, R)$ around the origin with radius R. Equation (2.1.1) yields the assertion. Moreover, as $\overset{\circ}{\operatorname{grad}}$ and $\overset{\circ}{\operatorname{div}}$ are clearly densely defined, their (negative) adjoints are linear operators (see Lemma B.4.5) and hence, so are their restrictions. ☐

Motivated by Proposition 2.1.3 we *define* the operators grad and div in $L^2(\Omega)$, $L^2(\Omega)^3$, respectively, as extensions of $\overset{\circ}{\text{grad}}$ and $\overset{\circ}{\text{div}}$ as follows:

$$\text{grad} := -\overset{\circ}{\text{div}}{}^* \quad \text{and} \quad \text{div} := -\overset{\circ}{\text{grad}}{}^*.$$

Remark 2.1.4 We mention here that $\text{dom}(\text{grad}) = H^1(\Omega)$ and $\text{dom}(\overset{\circ}{\text{grad}}) = H_0^1(\Omega)$ in a maybe somewhat more familiar notation. Consequently, we can use

$$\varphi \in \text{dom}\left(\overset{\circ}{\text{grad}}\right) \tag{2.1.2}$$

to encode the vanishing of φ at the boundary (Dirichlet boundary condition), whereas $\varphi \in \text{dom}(\text{grad})$ merely expresses weak differentiability but no boundary condition. Similarly,

$$\Psi \in \text{dom}\left(\overset{\circ}{\text{div}}\right) \tag{2.1.3}$$

encodes[2] the classical boundary condition of Ψ being tangential on the boundary (Neumann boundary condition), whereas

$$\Psi \in \text{dom}(\text{div}) = \{\Phi \in L^2(\Omega)^3;\ \text{div}\,\Phi \in L^2(\Omega)\}$$

---

[2]This is motivated by the classical Gauss theorem for Ω with smooth boundary and unit outward normal field $n: \partial\Omega \to \mathbb{R}^3$ with $|n| = 1$:

$$\langle \text{grad}\,\varphi | \Psi \rangle_{L^2(\Omega)^3} + \langle \varphi | \text{div}\,\Psi \rangle_{L^2(\Omega)}$$

$$= \int_{\partial\Omega} (\varphi(x)\,n_1(x)\,\Psi_1(x) + \varphi(x)\,n_2(x)\,\Psi_2(x) + \varphi(x)\,n_3(x)\,\Psi_3(x))\,do(x),$$

where now φ having no boundary constraints implies $n \cdot \Psi = 0$, if the right-hand side is supposed to vanish. Comparing in this case $\langle \text{grad}\,\varphi | \Psi \rangle_{L^2(\Omega)^3} + \langle \varphi | \text{div}\,\Psi \rangle_{L^2(\Omega)} = 0$ with the definition of adjoints suggests

$$\Psi \in \text{dom}\left(\text{grad}^*\right)$$

and

$$\text{grad}^*\,\Psi = -\,\text{div}\,\Psi.$$

But

$$\text{grad}^* = -\overset{\circ}{\text{div}}{}^{**} = -\overset{\circ}{\text{div}}$$

and so in particular

$$\Psi \in \text{dom}\left(\overset{\circ}{\text{div}}\right).$$

expresses the existence of a weak L^2-divergence but requires no boundary condition. Note that even if classical evaluation at the boundary does not make sense, we may still take (2.1.2) and (2.1.3) as generalized[3] Dirichlet or Neumann boundary conditions. We emphasize that we do *not* require *any* regularity of the boundary. In particular, open sets with fractal boundaries are very well admitted in our considerations up to this point.

The operator A in (2.0.2) in the Dirichlet case would be given by setting $C = \overset{\circ}{\mathrm{grad}}$ and thus

$$A = \begin{pmatrix} 0 & \mathrm{div} \\ \overset{\circ}{\mathrm{grad}} & 0 \end{pmatrix}. \tag{2.1.4}$$

In the Neumann case we would take $C = \mathrm{grad}$ leading to

$$A = \begin{pmatrix} 0 & \overset{\circ}{\mathrm{div}} \\ \mathrm{grad} & 0 \end{pmatrix}. \tag{2.1.5}$$

For sake of definiteness let us focus on the Neumann case (the Dirichlet case being analogous). Now various phenomena can be described by the choice of the material law operators M_0 and M_1.

2.1.1 The Classical Heat Equation

We get—in the right interpretation—the equations of heat diffusion (see [8, Lecture 7] for a derivation):

For this, let $\mu = \mu^* \in \mathcal{B}(L^2(\Omega))$, $\kappa \in \mathcal{B}(L^2(\Omega)^3)$. We assume there exists $d_1, d_2 \in \,]0, \infty[$ such that for all $\varphi \in L^2(\Omega)$ and $\Phi \in L^2(\Omega)^3$

$$\langle \mu\varphi | \varphi \rangle_{L^2(\Omega)} \geq d_1 \langle \varphi | \varphi \rangle_{L^2(\Omega)}, \text{ and } \langle \kappa\Phi | \Phi \rangle_{L^2(\Omega)^3} \geq d_2 \langle \Phi | \Phi \rangle_{L^2(\Omega)^3}. \tag{2.1.6}$$

We find:

Proposition 2.1.5 *Given μ, κ as introduced above and satisfying (2.1.6), we have that*

$$M_0 = \begin{pmatrix} \mu & 0 \\ 0 & 0 \end{pmatrix}, \quad M_1 = \begin{pmatrix} 0 & 0 \\ 0 & \kappa^{-1} \end{pmatrix} \tag{2.1.7}$$

satisfy (1.3.10) and (1.3.11) for some $c_1, c_2 \in \,]0, \infty[$.

---

[3]To invoke boundary trace results to formulate boundary value problems—as is commonly done in the literature—is unnecessary and frequently a distraction from the main issues of interest.

Proof Clearly M_0 is selfadjoint since μ is. Moreover, $M_0[H] = L^2(\Omega) \oplus \{0\} \subseteq L^2(\Omega) \oplus L^2(\Omega)^3$ and $[\{0\}]M_0 = \{0\} \oplus L^2(\Omega)^3 \subseteq L^2(\Omega) \oplus L^2(\Omega)^3$. Thus, the assertion follows from (2.1.6) since this implies that κ is continuously invertible and $\langle \kappa^{-1}\Phi | \Phi \rangle_{L^2(\Omega)^3} \geq (d_2/\|\kappa\|^2) \langle \Phi | \Phi \rangle_{L^2(\Omega)^3}$ for all $\Phi \in L^2(\Omega)^3$.   □

With this proposition at hand, we appeal to Proposition 1.3.9 and Theorem 1.3.2 and obtain well-posedness and causality of the equations describing heat diffusion:

Theorem 2.1.6 *Given μ, κ as above and satisfying (2.1.6), there exists $\varrho_0 \in {]}0, \infty[$ such that for all $\varrho \in [\varrho_0, \infty[$ the (closure of the) operator*

$$\left(\partial_0 \begin{pmatrix} \mu & 0 \\ 0 & 0 \end{pmatrix} + \begin{pmatrix} 0 & 0 \\ 0 & \kappa^{-1} \end{pmatrix} + \begin{pmatrix} 0 & \operatorname{div} \\ \operatorname{grad} & 0 \end{pmatrix} \right)$$

is continuously invertible in $H_{\varrho,0}(\mathbb{R}; L^2(\Omega)^{1+3})$. The inverse operator is causal and does not depend on ϱ (see Theorem 1.3.6 for the precise statement).

Remark 2.1.7 A solution (u_0, u_1) of

$$\left(\partial_0 \begin{pmatrix} \mu & 0 \\ 0 & 0 \end{pmatrix} + \begin{pmatrix} 0 & 0 \\ 0 & \kappa^{-1} \end{pmatrix} + \begin{pmatrix} 0 & \operatorname{div} \\ \operatorname{grad} & 0 \end{pmatrix} \right) \begin{pmatrix} u_0 \\ u_1 \end{pmatrix} = \begin{pmatrix} f_0 \\ f_1 \end{pmatrix}$$

for some given (f_0, f_1) in fact leads to a solution of the classical heat equation. Line by line we have

$$\partial_0 \mu u_0 + \operatorname{div} u_1 = f_0,$$

$$\kappa^{-1} u_1 + \operatorname{grad} u_0 = f_1.$$

If we put $f_1 = 0$ and interpret u_0 as temperature, u_1 as heat flux, and f_0 as a heat source density, we get Fourier's law

$$u_1 = -\kappa \operatorname{grad} u_0 \qquad\qquad (2.1.8)$$

and substituting this into the first equation, we obtain the classical heat equation with homogeneous Neumann boundary conditions

$$\partial_0 \mu u_0 - \operatorname{div}(\kappa \operatorname{grad} u_0) = f_0,$$

with μ being the volumetric heat capacity, that is, the product of mass density and specific heat capacity, and κ describing the heat conductivity. We note here that every step in this arguably formal procedure (we did not consider any operator domains here) can be made rigorous. Moreover, starting out with the classical formulation of the heat equation, we can reverse our argument to yield a solution of the original system. With the uniqueness statement at hand, we have well-posedness of the classical heat equation following from our general theory for evolutionary equations.

In fact, for the heat equation, it is possible to derive an adapted regularity statement using the framework of evolutionary equations, see [71].

2.1.2 The Maxwell–Cattaneo-Vernotte Model

A slight variant of the classical heat equation is the so-called Maxwell–Cattaneo–Vernotte (MCV) model of heat propagation, see e.g. [84]. The model proposes a modification of the Fourier law, namely

$$\partial_0 \tau u_1 + u_1 = -\kappa \operatorname{grad} u_0, \tag{2.1.9}$$

where τ is referred to as relaxation time, which we assume to be a fixed positive number. This amounts to consider

$$M_0 = \begin{pmatrix} \mu & 0 \\ 0 & \tau\kappa^{-1} \end{pmatrix}, \ M_1 = \begin{pmatrix} 0 & 0 \\ 0 & \kappa^{-1} \end{pmatrix}. \tag{2.1.10}$$

We again obtain the well-posedness of the corresponding system, however, since κ now occurs also in M_0, we need to assume the selfadjointness of κ as well:

Theorem 2.1.8 *Let μ and κ be as above satisfying (2.1.6) and assume additionally that $\kappa = \kappa^*$. Let $\tau \in \ ]0, \infty[$. Then M_0 and M_1 as in (2.1.10) satisfy (1.3.10) and (1.3.11) and the operator*

$$\overline{\left(\partial_0 \begin{pmatrix} \mu & 0 \\ 0 & \tau\kappa^{-1} \end{pmatrix} + \begin{pmatrix} 0 & 0 \\ 0 & \kappa^{-1} \end{pmatrix} + \begin{pmatrix} 0 & \operatorname{div} \\ \operatorname{grad} & 0 \end{pmatrix} \right)}^{-1} \in \mathcal{B}(H_{\varrho,0}(\mathbb{R}; L^2(\Omega)^4))$$

is well-defined, bounded, causal and independent of all sufficiently large $\varrho > 0$.

Remark 2.1.9 By a change of interpretation the resulting system also describes the linearized propagation of acoustic waves: Let f_0 be given and let (u_0, u_1) solve

$$\left(\partial_0 \begin{pmatrix} \mu & 0 \\ 0 & \tau\kappa^{-1} \end{pmatrix} + \begin{pmatrix} 0 & 0 \\ 0 & \kappa^{-1} \end{pmatrix} + \begin{pmatrix} 0 & \operatorname{div} \\ \operatorname{grad} & 0 \end{pmatrix} \right) \begin{pmatrix} u_0 \\ u_1 \end{pmatrix} = \begin{pmatrix} f_0 \\ 0 \end{pmatrix}. \tag{2.1.11}$$

Now, we interpret u_0 as pressure and $-u_1$ as the velocity field. Indeed, substituting u_1 from (2.1.9) (or the second equation in (2.1.11)) yields

$$\partial_0 \mu u_0 - \overset{\circ}{\operatorname{div}} (\partial_0 \tau + 1)^{-1} \kappa \operatorname{grad} u_0 = f_0.$$

Formally applying the operator $\partial_0 \tau + 1$ to both sides gives

$$\partial_0^2 \tau \mu u_0 + \partial_0 \mu u_0 - \operatorname{div} \kappa \operatorname{grad} u_0 = (\partial_0 \tau + 1) f_0 =: g_0, \qquad (2.1.12)$$

which is a damped wave equation for the pressure field.

The slightly more general situation

$$M_0 = \begin{pmatrix} \mu & 0 \\ 0 & \alpha \end{pmatrix}, \ M_1 = \begin{pmatrix} \beta & 0 \\ 0 & \gamma \end{pmatrix} \qquad (2.1.13)$$

results in a problem pertaining to waves in inhomogeneous anisotropic media. For if (u_0, u_1) solves

$$\left(\partial_0 \begin{pmatrix} \mu & 0 \\ 0 & \alpha \end{pmatrix} + \begin{pmatrix} \beta & 0 \\ 0 & \gamma \end{pmatrix} + \begin{pmatrix} 0 & \overset{\circ}{\operatorname{div}} \\ \operatorname{grad} & 0 \end{pmatrix} \right) \begin{pmatrix} u_0 \\ u_1 \end{pmatrix} = \begin{pmatrix} f_0 \\ 0 \end{pmatrix}$$

for some given f_0, substituting u_1 from the second equation into the first yields

$$\partial_0 \mu u_0 + \beta u_0 + \overset{\circ}{\operatorname{div}} (\partial_0 \alpha + \gamma)^{-1} \operatorname{grad} u_0 = f_0,$$

which requires restrictive assumptions on α, γ to be turned into a second order problem. This indicates that the first order approach is more general, allowing for more complicated mathematical models.

Remark 2.1.10 Systems with materials of the form (2.1.10) also appear as subsystems in the description of plasma and the propagation of holes in a porous medium, see e.g. [28].

2.2   A Reduction Mechanism and the Relativistic Schrödinger Equation

Sometimes it is useful to reduce a given problem to the range and null space of the operator A separately. This is an abstract procedure, which we therefore describe in general terms. This reduction procedure has been successfully applied to homogenization problems, see [85, 89]. For an application to elliptic type problems, we refer to [82].

2.2.1   Unitary Congruent Evolutionary Problems

We reconsider a general evolutionary operator

$$\partial_0 M_0 + M_1 + A \tag{2.2.1}$$

with a skew-selfadjoint operator A in some Hilbert space H. The reduction mechanism is based on the orthogonal decomposition of the underlying Hilbert space

$$H = \overline{A[H]} \oplus [\{0\}] A.$$

We introduce some notation:

Definition 2.2.1 Let H_0, H_1 be Hilbert spaces, $B: \operatorname{dom}(B) \subseteq H_0 \to H_1$ linear and closed. Then we define canonical embeddings

$$\iota_r(B): \overline{B[H_0]} \hookrightarrow H_1$$

$$\iota_n(B): [\{0\}]B \hookrightarrow H_0$$

from the (closure of the) range $\overline{\operatorname{ran}(B)} = \overline{B[H_0]}$ and the null space $\ker(B) = [\{0\}]B$ into H_1 and H_0, respectively. We shall also use

$$\iota_r^*(B) := \iota_r(B)^* \text{ and } \iota_n^*(B) := \iota_n(B)^*.$$

For a skew-selfadjoint operator A, we further define—in an obvious and suggestive use of block matrix notations—

$$\iota(A): \overline{A[H]} \oplus [\{0\}] A \to H$$

$$\begin{pmatrix} u \\ v \end{pmatrix} \mapsto \begin{pmatrix} \iota_r(A) & \iota_n(A) \end{pmatrix} \begin{pmatrix} u \\ v \end{pmatrix}$$

$$= \iota_r(A)u + \iota_n(A)v = u + v.$$

We note that $\iota(A)$ just defined is (obviously) a unitary operator. Furthermore, the adjoint/inverse of $\iota(A)$ is given by

$$\iota(A)^* : H \to \overline{A[H]} \oplus [\{0\}] A$$

$$x \mapsto \begin{pmatrix} \iota_r^*(A) \\ \iota_n^*(A) \end{pmatrix} x = \begin{pmatrix} \iota_r^*(A)x \\ \iota_n^*(A)x \end{pmatrix}.$$

We have the following observation:

Lemma 2.2.2 *Let A be a skew-selfadjoint operator in a Hilbert space H. Define $\widetilde{A} := \iota_{\mathrm{r}}(A)^* A\, \iota_{\mathrm{r}}(A)$. Then*

$$
\iota(A)^* A\, \iota(A) = \begin{pmatrix} \widetilde{A} & 0 \\ 0 & 0 \end{pmatrix}.
$$

Proof The result follows from $\iota_{\mathrm{n}}(A)^* A x = 0$ for all $x \in \mathrm{dom}\,(A)$ and $A\,\iota_{\mathrm{n}}(A) y = 0$ for all $y \in [\{0\}]\, A$. $\square$

Since $\iota(A)$ is unitary, we may write (2.2.1) in a unitarily congruent formulation

$$
\iota(A)^* \left(\partial_0 M_0 + M_1 + A\right) \iota(A) = \left(\partial_0 \widetilde{M}_0 + \widetilde{M}_1 + \begin{pmatrix} \widetilde{A} & 0 \\ 0 & 0 \end{pmatrix}\right),
$$

where

$$
\widetilde{M}_0 := \iota(A)^* M_0\, \iota(A) \quad\text{and}\quad \widetilde{M}_1 := \iota(A)^* M_1\, \iota(A).
$$

Unitary congruence preserves the structure of the system. In particular, the well-posedness constraints (1.3.1) for M_0 and M_1 are equivalent to the respective ones for $\widetilde{M}_0$ and $\widetilde{M}_1$. Thus, unitarily congruent systems are essentially the same. If the material law operators M_0 and M_1, are such that they commute with the orthogonal projector onto $\overline{A\,[H]}$ then we obtain a block diagonal system

$$
\partial_0 \begin{pmatrix} \widetilde{M}_{0,00} & 0 \\ 0 & \widetilde{M}_{0,11} \end{pmatrix} + \begin{pmatrix} \widetilde{M}_{1,00} & 0 \\ 0 & \widetilde{M}_{1,11} \end{pmatrix} + \begin{pmatrix} \widetilde{A} & 0 \\ 0 & 0 \end{pmatrix}
$$
$$
= \begin{pmatrix} \partial_0 \widetilde{M}_{0,00} + \widetilde{M}_{1,00} + \widetilde{A} & 0 \\ 0 & \partial_0 \widetilde{M}_{0,11} + \widetilde{M}_{1,11} \end{pmatrix}. \qquad (2.2.2)
$$

In this case the ordinary differential equation part (bottom right corner) decouples. Thus, inverting the operator

$$
\partial_0 M_0 + M_1 + A
$$

in the unitary equivalent form (2.2.2) reduces to solving an ordinary differential equation and a partial differential equation with a skew-selfadjoint operator $\widetilde{A}$ that has trivial kernel. Note that in applications $\widetilde{A}$ may even have compact resolvent, see, in particular, [89].

We shall now focus on a further abstract construction concerning the unbounded (spatial) part A, which will lead to a discussion of the relativistic Schrödinger equation:

Theorem 2.2.3 *Let*

$$A = \begin{pmatrix} 0 & -C^* \\ C & 0 \end{pmatrix}$$

with $C : \operatorname{dom}(C) \subseteq H_0 \to H_1$ *a closed, densely defined linear operator. Then, A is unitarily congruent to*

$$\begin{pmatrix} 0 & -|C| \\ |C| & 0 \end{pmatrix}$$

acting in $H_0 \oplus H_0$.

Proof We recall from Lemma 2.2.2 that

$$\begin{pmatrix} \widetilde{A} & 0 \\ 0 & 0 \end{pmatrix} = \iota(A)^* A \, \iota(A).$$

Since $\overline{A\,[H]} = \overline{C^*\,[H_1]} \oplus \overline{C\,[H_0]}$ we obtain

$$\iota_{\mathrm{r}}(A) = \begin{pmatrix} \iota_{\mathrm{r}}(C^*) & 0 \\ 0 & \iota_{\mathrm{r}}(C) \end{pmatrix} \text{ and } \iota_{\mathrm{r}}^*(A) = \begin{pmatrix} \iota_{\mathrm{r}}^*(C^*) & 0 \\ 0 & \iota_{\mathrm{r}}^*(C) \end{pmatrix}.$$

Thus,

$$\widetilde{A} = \iota_{\mathrm{r}}(A)^* A \, \iota_{\mathrm{r}}(A)$$

$$= \begin{pmatrix} \iota_{\mathrm{r}}^*(C^*) & 0 \\ 0 & \iota_{\mathrm{r}}^*(C) \end{pmatrix} \begin{pmatrix} 0 & -C^* \\ C & 0 \end{pmatrix} \begin{pmatrix} \iota_{\mathrm{r}}(C^*) & 0 \\ 0 & \iota_{\mathrm{r}}(C) \end{pmatrix}$$

$$= \begin{pmatrix} 0 & -\iota_{\mathrm{r}}^*(C^*)\,C^*\iota_{\mathrm{r}}(C) \\ \iota_{\mathrm{r}}^*(C)\,C\,\iota_{\mathrm{r}}(C^*) & 0 \end{pmatrix}$$

Next, we focus on $\iota_{\mathrm{r}}^*(C^*)\,C^*\,\iota_{\mathrm{r}}(C)$ and $\iota_{\mathrm{r}}^*(C)\,C\,\iota_{\mathrm{r}}(C^*) = \left(\iota_{\mathrm{r}}^*(C^*)\,C^*\iota_{\mathrm{r}}(C)\right)^*$. We define $X_0 := \overline{C^*[H_1]}$ and $X_1 := \overline{C[H_0]}$ and obtain that

$$\iota_{\mathrm{r}}^*(C)\,C\,\iota_{\mathrm{r}}\left(C^*\right) : \operatorname{dom}(C) \cap X_0 \subseteq X_0 \to X_1$$

is a densely defined closed linear operator with trivial null-space and dense range.

By the polar decomposition (see Proposition B.8.6) applied to this operator, we find a unitary operator

$$U : X_0 \to X_1$$

such that

$$\iota_r^* (C) \, C \, \iota_r \left(C^*\right) = U \left| \iota_r^* (C) \, C \, \iota_r \left(C^*\right) \right| .$$

Consequently,

$$\iota_r^* \left(C^*\right) C^* \iota_r (C) = \left(\iota_r^* (C) \, C \, \iota_r \left(C^*\right) \right)^*$$
$$= \left| \iota_r^* (C) \, C \, \iota_r \left(C^*\right) \right| U^* ,$$

which yields

$$\left(\iota_r^* \left(C^*\right) C^* \iota_r (C) \right) U = \left| \iota_r^* (C) \, C \, \iota_r \left(C^*\right) \right| .$$

Thus, we obtain the unitary congruence

$$\begin{pmatrix} 0 & - \left| \iota_r^* (C) \, C \, \iota_r (C^*) \right| \\ \left| \iota_r^* (C) \, C \, \iota_r (C^*) \right| & 0 \end{pmatrix}$$
$$= \begin{pmatrix} 1 & 0 \\ 0 & U^* \end{pmatrix} \begin{pmatrix} 0 & -\iota_r^* (C^*) \, C^* \iota_r (C) \\ \iota_r^* (C) \, C \, \iota_r (C^*) & 0 \end{pmatrix} \begin{pmatrix} 1 & 0 \\ 0 & U \end{pmatrix} .$$

Next, we use $\overline{|C| \, [H_0]} = \overline{C^* \, [H_1]} = X_0$ (see Eq. (B.8.4)). Thus, $\iota_r(C^*) = \iota_r(|C|)$ and

$$\left(\iota_r^*(|C|) \, |C| \, \iota_r(|C|) \right)^2 = \iota_r^*(|C|) \, |C| \, \iota_r(|C|) \, \iota_r^*(|C|) \, |C| \, \iota_r(|C|)$$
$$= \iota_r^*(|C|) \, |C|^2 \, \iota_r(|C|)$$
$$= \iota_r^*(C^*) C^* C \, \iota_r(C^*)$$
$$= \iota_r^*(C^*) C^* \, \iota_r(C) \, \iota_r^*(C) C \, \iota_r(C^*)$$
$$= \left(\left| \iota_r^*(C) C \, \iota_r(C^*) \right| \right)^2$$

and so, by uniqueness of roots (Theorem B.8.4) and since $\iota_r^*(|C|) \, |C| \, \iota_r(|C|)$ is accretive and selfadjoint, we obtain

$$\iota_r^*(|C|) \, |C| \, \iota_r(|C|) = \left| \iota_r^*(C) C \, \iota_r(C^*) \right| .$$

Thus we have shown that

$$
\left(\begin{pmatrix} 0 & -\left|\iota_r^*(C)C\,\iota_r(C^*)\right| \\ \left|\iota_r^*(C)C\,\iota_r(C^*)\right| & 0 \\ & 0 \end{pmatrix} \; 0 \atop 0 \right)
$$

$$
= \left(\begin{pmatrix} 0 & -\iota_r^*(|C|)\,|C|\,\iota_r(|C|) \\ \iota_r^*(|C|)\,|C|\,\iota_r(|C|) & 0 \\ & 0 \end{pmatrix} \; 0 \atop 0 \right)
$$

$$
= \iota \left(\begin{pmatrix} 0 & -|C| \\ |C| & 0 \end{pmatrix} \right)^* \left(\begin{pmatrix} 0 & -|C| \\ |C| & 0 \end{pmatrix} \iota \left(\begin{pmatrix} 0 & -|C| \\ |C| & 0 \end{pmatrix} \right) \right).
$$

Hence, we may now easily verify that

$$
A = W^* \begin{pmatrix} 0 & -|C| \\ |C| & 0 \end{pmatrix} W,
$$

where

$$
W = \iota \left(\begin{pmatrix} 0 & -|C| \\ |C| & 0 \end{pmatrix} \right) \left(\begin{pmatrix} 1 & 0 \\ 0 & U^* \end{pmatrix} \; 0 \atop 1 \right) \iota(A)^*,
$$

which is a composition of unitary operators and, thus, unitary. □

2.2.2 The Relativistic Schrödinger Equation

We return to considerations of our general evolutionary operator

$$
\partial_0 M_0 + M_1 + \begin{pmatrix} 0 & -C^* \\ C & 0 \end{pmatrix}.
$$

Theorem 2.2.3 states that there exists a unitary operator W such that

$$
W^* \left(\partial_0 M_0 + M_1 + \begin{pmatrix} 0 & -C^* \\ C & 0 \end{pmatrix} \right) W = \left(\partial_0 \tilde{M}_0 + \tilde{M}_1 + \begin{pmatrix} 0 & -|C| \\ |C| & 0 \end{pmatrix} \right),
$$

where

$$\tilde{M}_0 := W^* M_0 W \text{ and } \tilde{M}_1 := W^* M_1 W.$$

Thus, dropping the "~" to simplify notation, we consider evolutionary operators of the specific form

$$\partial_0 M_0 + M_1 + \begin{pmatrix} 0 & -|C| \\ |C| & 0 \end{pmatrix} \tag{2.2.3}$$

rather than of the general form (2.0.1) and (2.0.2). In particular, if the material law operators M_0, M_1 in (2.2.3) also commute with $i := \begin{pmatrix} 0 & -1 \\ 1 & 0 \end{pmatrix}$ we may consider (2.2.3) as an equation in the complexifaction of H_0 and write

$$\partial_0 M_0 + M_1 + i\,|C|\,. \tag{2.2.4}$$

Using (2.1.4) as an application of this procedure, we obtain

$$\partial_0 M_0 + M_1 + A \tag{2.2.5}$$

with

$$A = \begin{pmatrix} 0 & -\left|\overset{\circ}{\text{grad}}\right| \\ \left|\overset{\circ}{\text{grad}}\right| & 0 \end{pmatrix}. \tag{2.2.6}$$

This is a proper evolutionary problem in a complexified $L^2(\Omega)_{\mathbb{C}}$ since by assumption

$$\begin{pmatrix} 0 & -1 \\ 1 & 0 \end{pmatrix} M_0 = M_0 \begin{pmatrix} 0 & -1 \\ 1 & 0 \end{pmatrix}, \quad \begin{pmatrix} 0 & -1 \\ 1 & 0 \end{pmatrix} M_1 = M_1 \begin{pmatrix} 0 & -1 \\ 1 & 0 \end{pmatrix}, \tag{2.2.7}$$

in other words M_0 and M_1 are "real", in the sense that they commute with "taking the real-part". Since we restricted the original acoustic equations to real solutions, the underlying Hilbert space is actually based on real-valued function in $L^2(\Omega)$, that is, in $L^2(\Omega; \mathbb{R})$. From this perspective, the operator in (2.2.6) would be considered in $L^2(\Omega, \mathbb{R}) \oplus L^2(\Omega, \mathbb{R})$ and can be written in complex notation[4]

$$\partial_0 M_0 + M_1 + i\,\left|\overset{\circ}{\text{grad}}\right| \tag{2.2.8}$$

---

[4]The standard (non-relativistic) Schrödinger equation (for a homogeneous Dirichlet boundary condition) appears if $\left|\overset{\circ}{\text{grad}}\right|$ is replaced by $\left|\overset{\circ}{\text{grad}}\right|^2 = -\operatorname{div}\overset{\circ}{\text{grad}} \subseteq -\Delta$.

acting in the complexification $L^2(\Omega, \mathbb{C}) \equiv L^2(\Omega, \mathbb{R})_{\mathbb{C}}$ of $L^2(\Omega, \mathbb{R})$, which for our purposes will still be considered as a real Hilbert space (note that $\mathbb{C}$ is a Hilbert space over $\mathbb{R}$ with $(z, \zeta) \mapsto \mathfrak{Re}(\overline{z}\zeta)$ as inner product). In this perspective i coincides with the standard complex unit. The resulting complex equation (2.2.8) is discussed in the literature under the title "relativistic Schrödinger equation" (with mass zero, usually $M_0 = 1$ and M_1 is a scalar multiplication operator so that (2.2.7) is clearly satisfied).

Remark 2.2.4 Let us briefly consider the non-zero mass relativistic Schrödinger equation. Consider (2.2.5) with $M_0 = 1$ and $M_1 = \Theta + \begin{pmatrix} 0 & i \\ i & 0 \end{pmatrix} \mu$ for some $\mu \in [0, \infty[$ and a bounded linear operator Θ. Then

$$\partial_0 M_0 + M_1 + A = \partial_0 + \Theta + \begin{pmatrix} 0 & -\left|\overset{\circ}{\mathrm{grad}}\right| + i\mu \\ \left|\overset{\circ}{\mathrm{grad}}\right| + i\mu & 0 \end{pmatrix}. \tag{2.2.9}$$

To the operator

$$\begin{pmatrix} 0 & -\left|\overset{\circ}{\mathrm{grad}}\right| + i\mu \\ \left|\overset{\circ}{\mathrm{grad}}\right| + i\mu & 0 \end{pmatrix}$$

we may apply Theorem 2.2.3 (note that this result is not restricted to a real Hilbert space and applies verbatim to the complex case with literally the same proof). Thus, for appropriate Θ the operator in (2.2.9) is unitarily congruent to

$$\partial_0 + \widetilde{\Theta} + \begin{pmatrix} 0 & -\left|\left|\overset{\circ}{\mathrm{grad}}\right| + i\mu\right| \\ \left|\left|\overset{\circ}{\mathrm{grad}}\right| + i\mu\right| & 0 \end{pmatrix},$$

which in the notation of our abstract complexification introduced earlier is

$$\partial_0 + \widetilde{\Theta} + i \left|\left|\overset{\circ}{\mathrm{grad}}\right| + i\mu\right| = \partial_0 + V + i \left(\left|\left|\overset{\circ}{\mathrm{grad}}\right| + i\mu\right| - \mu\right)$$

with

$$V := \widetilde{\Theta} + i\mu.$$

The resulting equation, assuming V commuting with i, is

$$\left(\partial_0 + V + i \left(\left|\left|\overset{\circ}{\mathrm{grad}}\right| + i\mu\right| - \mu\right)\right)\varphi = f,$$

which is known as the "relativistic Schrödinger equation" with mass[5] μ. Of course, the particular interest in singular multipliers V is beyond the reach of our current assumptions, but in any case—up to unitary congruence—the issues arising are the same as in the "acoustic" case. The differences are only in the particular form of the material law operators.

Material laws of the form (2.1.13) describe typical classical materials. If non-block-diagonal entries and/or additional higher order (in ∂_0^{-1}) terms occur then the label "acoustic meta-materials" is sometimes used. Due to the general solution theory of Sect. 1.3 and Theorem 1.3.11 in particular, meta-materials are easily covered by our approach.

We shall postpone a discussion of some of the issues involved in connection with meta-materials until we discuss the analogous situation for electromagnetic phenomena.

2.3   Linear Elasticity

The operations grad and div can be generalized to tensors of higher order. We will use the same notation for the respective generalized operators as it will always be clear from the context, what the order of the tensors are. Furthermore, to keep matters simple we shall, as already done in the previous discussions, stay with the Cartesian setting and only consider gradients of vector fields and divergences of 2-tensors, which are in the Cartesian setting just square matrices, see [48, 52, 60] for a more general perspective. Gradients of vector fields $v = \begin{pmatrix} v_1 \\ v_2 \\ v_3 \end{pmatrix}$ are taken component-wise but arranged as rows

$$\operatorname{grad} v := \begin{pmatrix} (\operatorname{grad} v_1)^\top \\ (\operatorname{grad} v_2)^\top \\ (\operatorname{grad} v_3)^\top \end{pmatrix} = \begin{pmatrix} \partial_1 v_1 & \partial_2 v_1 & \partial_3 v_1 \\ \partial_1 v_2 & \partial_2 v_2 & \partial_3 v_2 \\ \partial_1 v_3 & \partial_2 v_3 & \partial_3 v_3 \end{pmatrix},$$

in other words, grad v is identified with the Jacobian of v as a mapping from $\Omega \subseteq \mathbb{R}^3$ to $\mathbb{R}^3$. Thus,

$$\operatorname{grad}: \operatorname{dom}(\operatorname{grad})^3 \subseteq L^2(\Omega)^3 \to L^2(\Omega)^{3 \times 3}.$$

---

[5]By taking $\left| \overset{\circ}{\operatorname{grad}} + i\mu \right|$ instead, with μ now being a vector field (acting as a bounded multiplier), we obtain the "magnetic" case, where μ is a vector potential of the magnetic field. Again in the standard non-relativistic case $\left| \overset{\circ}{\operatorname{grad}} + i\mu \right|^2$ is used in place of $\left| \overset{\circ}{\operatorname{grad}} + i\mu \right|$.

Likewise, divergence is extended to matrices $\alpha = \begin{pmatrix} \alpha_{11} & \alpha_{12} & \alpha_{13} \\ \alpha_{21} & \alpha_{22} & \alpha_{23} \\ \alpha_{31} & \alpha_{32} & \alpha_{33} \end{pmatrix} = \begin{pmatrix} \alpha_1 \\ \alpha_2 \\ \alpha_3 \end{pmatrix}$ with rows α_k,

$k \in \{1, 2, 3\}$, row-wise and arranged as a column:

$$\operatorname{div} \alpha = \begin{pmatrix} \operatorname{div} \alpha_1^\top \\ \operatorname{div} \alpha_2^\top \\ \operatorname{div} \alpha_3^\top \end{pmatrix} = \begin{pmatrix} \sum_{k=1}^{3} \partial_k \alpha_{1k} \\ \sum_{k=1}^{3} \partial_k \alpha_{2k} \\ \sum_{k=1}^{3} \partial_k \alpha_{3k} \end{pmatrix},$$

which eventually yields

$$\operatorname{div} \colon \operatorname{dom}(\operatorname{div})^3 \subseteq L^2(\Omega)^{3\times 3} \to L^2(\Omega)^3.$$

As before, we analogously adopt the notation $\overset{\circ}{\operatorname{grad}}$ and $\overset{\circ}{\operatorname{div}}$ for respective restrictions of grad and div discussed here to vector fields and matrices with (generalized) homogeneous Dirichlet and Neumann boundary conditions. Then as before, we define the operators div and grad as the skew-adjoints of $\overset{\circ}{\operatorname{grad}}$ and $\overset{\circ}{\operatorname{div}}$, respectively. Thus, we maintain, for example in the Dirichlet case, the skew-selfadjointness of

$$\begin{pmatrix} 0 & \operatorname{div} \\ \overset{\circ}{\operatorname{grad}} & 0 \end{pmatrix}$$

but now in the Hilbert space $L^2(\Omega)^3 \oplus L^2(\Omega)^{3\times 3}$, where $L^2(\Omega)^{3\times 3}$ is equipped with the natural inner product[6] induced by considering 3×3-matrices as a triple of row vectors, i.e., for $\alpha = \begin{pmatrix} \alpha_{11} & \alpha_{12} & \alpha_{13} \\ \alpha_{21} & \alpha_{22} & \alpha_{23} \\ \alpha_{31} & \alpha_{32} & \alpha_{33} \end{pmatrix} = \begin{pmatrix} \alpha_1 \\ \alpha_2 \\ \alpha_3 \end{pmatrix}$ and $\beta = \begin{pmatrix} \beta_{11} & \beta_{12} & \beta_{13} \\ \beta_{21} & \beta_{22} & \beta_{23} \\ \beta_{31} & \beta_{32} & \beta_{33} \end{pmatrix} = \begin{pmatrix} \beta_1 \\ \beta_2 \\ \beta_3 \end{pmatrix}$:

$$\langle \alpha | \beta \rangle_{L^2(\Omega)^{3\times 3}} = \sum_{k=1}^{3} \langle \alpha_k | \beta_k \rangle_{L^2(\Omega)^3}.$$

---

[6]Another way of expressing this inner product is by considering the matrix trace (trace $\alpha = \sum_{k=1}^{3} \alpha_{kk}$) applied to the matrix product $\alpha^* \beta$, that is, the so-called Frobenius inner product

$$\mathbb{R}^{3\times 3} \times \mathbb{R}^{3\times 3} \to \mathbb{R}$$

$$(\alpha, \beta) \mapsto \operatorname{trace}\left(\alpha^* \beta\right)$$

and lifting it to the matrix-valued case

$$L^2(\Omega)^{3\times 3} \times L^2(\Omega)^{3\times 3} \to \mathbb{R}$$

$$(\alpha, \beta) \mapsto \int_\Omega \operatorname{trace}\left(\alpha(x)^* \beta(x)\right) dx.$$

2.3.1 General (Non-symmetric) Linear(ized) Elasticity

We shall focus on the Dirichlet boundary condition skipping the obvious variant for the Neumann case. We consider operators of the form

$$\partial_0 M_0 + M_1 + \begin{pmatrix} 0 & \operatorname{div} \\ \overset{\circ}{\operatorname{grad}} & 0 \end{pmatrix} \tag{2.3.1}$$

in $H = L^2(\Omega)^3 \oplus L^2(\Omega)^{3 \times 3}$. In the context of (non-symmetric) linear elasticity theory (see e.g. [41, 42]) for classical materials, M_0, M_1 would be block diagonal corresponding to the block decomposition given by the operator containing the spatial derivatives. For example, letting

$$M_0 = \begin{pmatrix} \mu_* & 0 \\ 0 & C^{-1} \end{pmatrix}, \quad M_1 = 0 \tag{2.3.2}$$

would lead to an equation of the typical form

$$\left(\partial_0 \begin{pmatrix} \mu_* & 0 \\ 0 & C^{-1} \end{pmatrix} + \begin{pmatrix} 0 & \operatorname{div} \\ \overset{\circ}{\operatorname{grad}} & 0 \end{pmatrix} \right) \begin{pmatrix} v \\ -T \end{pmatrix} = \begin{pmatrix} f \\ g \end{pmatrix}, \tag{2.3.3}$$

where, to apply our general solution theory, we would assume that

$$\mu_* = \mu_*^*, \text{ and } C = C^* \text{ are strictly positive definite}$$

in their respective component spaces. It is easy to see that, so defined, M_0 and M_1 satisfy (1.3.1) and, thus, Theorem 1.3.2 applies. We will not bother to state the corresponding well-posedness theorem but instead consider an application of this system.

Equation (2.3.3) block-component-wise with $g = 0$, gives

$$\partial_0 \mu_* v - \operatorname{div} T = f,$$

$$\partial_0 T = C \operatorname{grad} v.$$

In the context of elasticity theory (see again [41, 42], compare also [4]): μ_* denotes mass density, v the velocity field of the deformation process, T the elastic stress and C describes elastic material properties. Re-writing this in terms of the elastic displacement

$$u := \partial_0^{-1} v$$

we get

$$\partial_0 \mu_* \partial_0 u - \operatorname{div} T = f \tag{2.3.4}$$

and the so-called Hooke's law

$$T = \mathcal{C} \operatorname{grad} u. \tag{2.3.5}$$

Substituting (2.3.5) into (2.3.4), we get the more commonly known formulation of linear elasticity as a second order equation in terms of the displacement u:

$$\partial_0 \mu_* \partial_0 u - \operatorname{div} \mathcal{C} \operatorname{grad} u = f. \tag{2.3.6}$$

Equation (2.3.3) (or (2.3.6)) describes the dynamics of what is known as non-symmetric linear elasticity.

2.3.2   The Isotropic Case

Often of particular interest is the isotropic case. For this we need to introduce some operators:

Definition 2.3.1 Let $\Omega \subseteq \mathbb{R}^3$ open. We define the following closed subspaces of $L^2(\Omega)^{3 \times 3}$:

$$L^2_{\text{sym}}(\Omega) := \left\{ T \in L^2(\Omega)^{3 \times 3} \mid T^\top = T \text{ a.e.} \right\},$$

and

$$L^2_{\text{skew}}(\Omega) := \left\{ T \in L^2(\Omega)^{3 \times 3} \mid T^\top = -T \text{ a.e.} \right\},$$

where the transposition $T \mapsto T^\top$ is to be understood as taken point-wise. We introduce the canonical embeddings

$$\iota_{\text{sym}} \colon L^2_{\text{sym}}(\Omega) \hookrightarrow L^2(\Omega)^{3 \times 3} \text{ and } \iota_{\text{skew}} \colon L^2_{\text{skew}}(\Omega) \hookrightarrow L^2(\Omega)^{3 \times 3},$$

and the trace operator

$$\text{trace} \colon L^2(\Omega)^{3 \times 3} \to L^2(\Omega), \, \left(\varphi_{ij} \right)_{i,j \in \{1,2,3\}} \mapsto \sum_{j=1}^{3} \varphi_{jj}.$$

Proposition 2.3.2 *The adjoint operators ι_{sym}^*, ι_{skew}^*, and trace* are given by*

$$\iota_{sym}^* T = \frac{1}{2}\left(T + T^\top\right), \quad \iota_{skew}^* T = \frac{1}{2}\left(T - T^\top\right), \quad \text{trace}^* p = \begin{pmatrix} p & 0 & 0 \\ 0 & p & 0 \\ 0 & 0 & p \end{pmatrix}$$

for all $T \in L^2(\Omega)^{3\times3}$ and $p \in L^2(\Omega)$ and the operators[7]

$$\text{sym} := \iota_{sym}\iota_{sym}^*, \quad \text{skew} := \iota_{skew}\iota_{skew}^*, \quad \text{trp} := \frac{1}{3}\text{trace}^* \text{trace}$$

are orthogonal projectors in $L^2(\Omega)^{3\times3}$.

   Moreover,

$$\text{sym}_0 := \text{sym} - \text{trp}$$

is also an orthogonal projection.

Proof We confine ourselves to a proof of the formula for trace*. This follows from the matrix calculation

$$\langle \text{trace}^* p | T \rangle_{L^2(\Omega)^{3\times3}} = \langle p | \text{trace}\, T \rangle_{L^2(\Omega)}$$

$$= \int_\Omega p\, \text{trace}\, T$$

---

[7] The mapping $A \mapsto \frac{1}{3}\text{trace}^*\text{trace}\, A = \text{trp}\, A$ is the orthogonal projector onto the direction of the metric tensor g, which in our case is just the identity matrix $g = 1_{\mathbb{R}^3}$. Indeed, in the Cartesian case

$$\text{trace}\, A = \left(1_{\mathbb{R}^3} | A\right)_{\mathbb{R}^{3\times3}}$$

and

$$\text{trp}\, A = \left\langle \frac{1}{|1_{\mathbb{R}^3}|_{\mathbb{R}^{3\times3}}} 1_{\mathbb{R}^3} \Big| A \right\rangle_{\mathbb{R}^{3\times3}} \frac{1}{|1_{\mathbb{R}^3}|_{\mathbb{R}^{3\times3}}} 1_{\mathbb{R}^3}.$$

Note that

$$|1_{\mathbb{R}^3}|_{\mathbb{R}^{3\times3}} = \sqrt{3}.$$

$$= \sum_{i=1}^{3} \int_{\Omega} p T_{ii}$$

$$= \left\langle \begin{pmatrix} p & 0 & 0 \\ 0 & p & 0 \\ 0 & 0 & p \end{pmatrix} \middle| T \right\rangle_{L^2(\Omega)^{3\times 3}},$$

for all $p \in L^2(\Omega)$ and $T \in L^2(\Omega)^{3\times 3}$.

Thus, for all $p \in L^2(\Omega)$ we obtain

$$\text{trace trace}^* p = \text{trace} \begin{pmatrix} p & 0 & 0 \\ 0 & p & 0 \\ 0 & 0 & p \end{pmatrix} = 3p$$

and hence,

$$\text{trp} := \frac{1}{3}\text{trace}^*\text{trace} = \frac{1}{9}\text{trace}^*\text{trace trace}^*\text{trace},$$

which shows the idempotence of trp. Hence, since trp is selfadjoint, it is an orthogonal projection.

The last assertion follows from

$$\text{trp} = \text{sym trp} = \text{trp sym.} \qquad \qquad \square$$

Corollary 2.3.3 *We have*[8]

$$\text{sym}_0 + \text{skew} + \text{trp} = 1$$

----

[8]The associated orthogonal projections of a tensor field $T \in L^2(\Omega)^{3\times 3}$ are referred to as deviatoric, rotational and volumetric (also mean or spherical) part, respectively. The term "mean" tensor field results from $\frac{1}{3}\text{trace } A$ being the average of the diagonal entries of A. The fact that the level surfaces of the quadratic form associated with $1_{\mathbb{R}^3}$, compare Footnote 7, are spheres (for a general g ellipsoids) motivates the term "spherical" tensor fields . The term "volumetric" stems from the interpretation of the trace of the strain in elasticity theory as the (approximate) relative volume change. By definition

$$\text{sym}_0 = \text{sym} - \text{trp}$$

and so sym_0 shows the deviation of the symmetric part from the mean tensor field, hence "deviatoric". Finally, since skew-symmetric tensors are generators of rotations, skew A is referred to as the "rotational" part of A.

and

$$L^2(\Omega)^{3\times3} = \mathrm{sym}_0\left[L^2(\Omega)^{3\times3}\right] \oplus \mathrm{skew}\left[L^2(\Omega)^{3\times3}\right] \oplus \mathrm{trp}\left[L^2(\Omega)^{3\times3}\right].$$

Proof Since skew $= 1 - \mathrm{sym}$, the first equality holds. The orthogonal decomposition is a consequence of the equality just proved as the orthogonal projections annihilate one another, see also Proposition 2.3.2. Indeed, it is easy to carry over the result

$$\mathbb{R}^{3\times3} = \mathrm{sym}_0\left[\mathbb{R}^{3\times3}\right] \oplus \mathrm{skew}\left[\mathbb{R}^{3\times3}\right] \oplus \mathrm{trp}\left[\mathbb{R}^{3\times3}\right]$$

to the $\mathbb{R}^{3\times3}$-valued case. $\square$

In the isotropic case the material dependent operator $\mathcal{C}$ is determined by three real parameters c_0, c_1, and c_2 via

$$\mathcal{C} = c_0\mathrm{sym} + c_1\mathrm{skew} + c_2\,\mathrm{trp} \tag{2.3.7}$$
$$= 2\mu\mathrm{sym}_0 + c_1\mathrm{skew} + (3\lambda + 2\mu)\,\mathrm{trp},$$

where $\lambda := \frac{c_2}{3}$ and $\mu := \frac{c_0}{2}$ are the so-called Lamé constants.

As a consequence of the latter observations, we find that $\mathcal{C}$ is strictly positive definite if and only if

$$c_0 > 0, \quad c_1 > 0, \quad \text{and } c_2 + c_0 > 0,$$

which is equivalent to $c_1 > 0$ and the familiar relations

$$\mu, \; 2\mu + 3\lambda > 0.$$

2.3.3   Symmetric Stresses

It is, however, usually assumed that the material properties encoded in the coefficient $\mathcal{C}$ are such that the stress is symmetric. Thus, in order to maintain selfadjointness of $\mathcal{C}$ as a mapping into the symmetric elements from $\mathrm{sym}\left[L^2(\Omega)^{3\times3}\right]$ into itself, we need to ask for

$$\mathcal{C}\,\mathrm{skew} = \mathrm{skew}\,\mathcal{C} = 0.$$

In this case, the mother-and-descendant mechanism of Sect. A.1 comes into play. We consider instead

$$\partial_0 M_0 + M_1 + \begin{pmatrix} 0 & \operatorname{Div} \\ \overset{\circ}{\operatorname{Grad}} & 0 \end{pmatrix},$$

where

$$\begin{pmatrix} 0 & \operatorname{Div} \\ \overset{\circ}{\operatorname{Grad}} & 0 \end{pmatrix} := \overline{\begin{pmatrix} 1 & 0 \\ 0 & \iota^*_{\mathrm{sym}} \end{pmatrix} \begin{pmatrix} 0 & \operatorname{div} \\ \overset{\circ}{\operatorname{grad}} & 0 \end{pmatrix} \begin{pmatrix} 1 & 0 \\ 0 & \iota_{\mathrm{sym}} \end{pmatrix}}$$

$$= \begin{pmatrix} 0 & \operatorname{div} \iota_{\mathrm{sym}} \\ \overline{\iota^*_{\mathrm{sym}} \overset{\circ}{\operatorname{grad}}} & 0 \end{pmatrix}$$

in the underlying Hilbert space $H := L^2(\Omega)^3 \oplus \operatorname{sym}[L^2(\Omega)^{3\times3}]$; note that as product of a closed and a bounded operator, the operator $\operatorname{div} \iota_{\mathrm{sym}}$ is already closed. That is, by definition

$$\overset{\circ}{\operatorname{Grad}} = \overline{\iota^*_{\mathrm{sym}} \overset{\circ}{\operatorname{grad}}}$$

$$\operatorname{Div} = \operatorname{div} \iota_{\mathrm{sym}}$$

and it is easy to show that

$$\operatorname{Div} = -\overset{\circ}{\operatorname{Grad}}{}^*.$$

Here we have—for notational simplicity—written M_0 and M_1 in place of $\begin{pmatrix} 1 & 0 \\ 0 & \iota^*_{\mathrm{sym}} \end{pmatrix} M_0 \begin{pmatrix} 1 & 0 \\ 0 & \iota_{\mathrm{sym}} \end{pmatrix}$ and $\begin{pmatrix} 1 & 0 \\ 0 & \iota^*_{\mathrm{sym}} \end{pmatrix} M_1 \begin{pmatrix} 1 & 0 \\ 0 & \iota_{\mathrm{sym}} \end{pmatrix}$, respectively. In the case of classical symmetric linear elasticity, the material law operators (2.3.2) are $M_1 = 0$ and

$$M_0 = \begin{pmatrix} \mu_* & 0 \\ 0 & \left(\iota^*_{\mathrm{sym}} C \iota_{\mathrm{sym}} \right)^{-1} \end{pmatrix}$$

for which we shall write again simply as

$$M_0 = \begin{pmatrix} \mu_* & 0 \\ 0 & C^{-1} \end{pmatrix}$$

with the understanding that the elasticity coefficient C is a continuous, selfadjoint, strictly positive definite mapping in $\operatorname{sym}[L^2(\Omega)^{3\times3}]$. In the isotropic case, that is, (2.3.7) with

$c_1 = 0$, we now have

$$\mathcal{C} = c_0 \iota^*_{\text{sym}} \text{sym}_0 \, \iota_{\text{sym}} + (c_2 + c_0) \, \iota^*_{\text{sym}} \, \text{trp} \, \iota_{\text{sym}},$$

$$= 2\mu \iota^*_{\text{sym}} \text{sym}_0 \, \iota_{\text{sym}} + \left(\frac{2}{3}\mu + \lambda \right) \iota^*_{\text{sym}} \, \text{trace}^* \text{trace} \, \iota_{\text{sym}}.$$

In the remaining part of this section, we shall compute $\overline{\iota^*_{\text{sym}}\text{grad}}$. It turns out that

$$\overline{\iota^*_{\text{sym}}\text{grad}} = \iota^*_{\text{sym}}\overset{\circ}{\text{grad}}.$$

The latter result is a consequence of a variant of Korn's first inequality:

Theorem 2.3.4 *Let $\Omega \subseteq \mathbb{R}^3$ be open. Then, for all $v \in \text{dom}(\overset{\circ}{\text{grad}})$*

$$\left| \overset{\circ}{\text{grad}} \, v \right|_{L^2(\Omega)^{3\times3}} \leq \sqrt{2} \left| \iota^*_{\text{sym}}\overset{\circ}{\text{grad}} \, v \right|_{\text{sym}[L^2(\Omega)^{3\times3}]}. \tag{2.3.8}$$

Proof We calculate, for $v \in \overset{\circ}{C}_\infty(\Omega)^3$, using integration by parts:

$$\left| \iota^*_{\text{sym}}\overset{\circ}{\text{grad}} \, v \right|^2_{\text{sym}[L^2(\Omega)^{3\times3}]} = \frac{1}{4} \sum_{i,j=1}^3 \int_\Omega \left((\partial_i v_j)(x) + (\partial_j v_i)(x) \right)^2 \, dx$$

$$= \frac{1}{2} \sum_{i,j=1}^3 \left(\int_\Omega (\partial_i v_j)(x)^2 \, dx + \int_\Omega (\partial_i v_j)(x)(\partial_j v_i)(x) \, dx \right)$$

$$= \frac{1}{2} \sum_{i,j=1}^3 \left(\int_\Omega (\partial_i v_j)(x)^2 \, dx + \int_\Omega (\partial_j v_j)(x)(\partial_i v_i)(x) \, dx \right)$$

$$= \frac{1}{2} \left| \overset{\circ}{\text{grad}} \, v \right|^2_{L^2(\Omega)^{3\times3}} + \frac{1}{2} \left| \overset{\circ}{\text{div}} \, v \right|^2_{L^2(\Omega)}$$

$$\geq \frac{1}{2} \left| \overset{\circ}{\text{grad}} \, v \right|^2_{L^2(\Omega)^{3\times3}}.$$

Thus, we obtain the inequality for the operator core $\overset{\circ}{C}_\infty(\Omega)^3$ of $\overset{\circ}{\text{grad}}$, which yields the assertion. □

Remark 2.3.5 The classical Korn's first inequality (see [15, 24, 25]) actually is a combination of (2.3.8) and the Poincaré estimate lifted to the vector field case:

$$|v|_{L^2(\Omega)^3} \leq c_P \left| \overset{\circ}{\text{grad}} \, v \right|_{L^2(\Omega)^{3\times3}}, \tag{2.3.9}$$

which, noting that (2.3.9) implies

$$|v|_{\text{dom}(\mathring{\text{grad}})} = \sqrt{|v|^2_{L^2(\Omega)^3} + \left|\mathring{\text{grad}}\, v\right|^2_{L^2(\Omega)^{3\times3}}} \leq \sqrt{1 + c_P^2}\, \left|\mathring{\text{grad}}\, v\right|_{L^2(\Omega)^{3\times3}},$$

yields with (2.3.8) the classical Korn's first inequality

$$|v|_{\text{dom}(\mathring{\text{grad}})} \leq \sqrt{2\left(1 + c_P^2\right)}\, \left|\iota^*_{\text{sym}}\mathring{\text{grad}}\, v\right|_{L^2(\Omega)^{3\times3}}. \tag{2.3.10}$$

Corollary 2.3.6 *Let $\Omega \subseteq \mathbb{R}^3$ be open. Then*

$$\iota^*_{\text{sym}}\mathring{\text{grad}} : \text{dom}(\mathring{\text{grad}}) \subseteq L^2(\Omega)^3 \to L^2_{\text{sym}}(\Omega)$$

is closed, that is,

$$\overline{\iota^*_{\text{sym}}\mathring{\text{grad}}} = \iota^*_{\text{sym}}\mathring{\text{grad}}.$$

Proof Let $(v_n)_n$ be a sequence in $\text{dom}(\mathring{\text{grad}})$ such that $v_n \to v \in L^2(\Omega)^3$ and $\iota^*_{\text{sym}}\mathring{\text{grad}}\, v_n \to w$ in $\text{sym}[L^2(\Omega)^{3\times3}]$ as $n \to \infty$ for some $v \in L^2(\Omega)^3$ and $w \in \text{sym}[L^2(\Omega)^{3\times3}]$. By Theorem 2.3.4 we deduce that $\left(\mathring{\text{grad}}\, v_n\right)_n$ is a Cauchy sequence in $L^2(\Omega)^{3\times3}$. Let $\tilde{w}$ be its limit. Due to the closedness of $\mathring{\text{grad}}$, we obtain $v \in \text{dom}(\mathring{\text{grad}})$ and $\mathring{\text{grad}}\, v = \tilde{w}$. By the continuity of ι^*_{sym}, we deduce that $v \in \text{dom}(\iota^*_{\text{sym}}\mathring{\text{grad}})$ and $\iota^*_{\text{sym}}\mathring{\text{grad}}\, v = \iota^*_{\text{sym}}\tilde{w} = w$, by the uniqueness of the limit. This proves the assertion. $\square$

2.3.4 Linearized Incompressible Stokes Equations

In the previous section, we met the mother-and-descendant mechanism (Sect. A.1) for the first time. In the present section, we will comment on a certain "non-example" of this, see also [56]. Consider again operators of the type

$$\partial_0 M_0 + M_1 + A$$

with

$$A = \begin{pmatrix} 0 & -C^* \\ C & 0 \end{pmatrix}$$

for some densely defined closed linear operators $C: \text{dom}(C) \subseteq H_0 \to H_1$, see also (2.0.1) and (2.0.2). For the linearized incompressible Navier–Stokes equations, that is, the Stokes and the (linear) Oseen equations, the operator A given by $\begin{pmatrix} 0 & \mathrm{Div} \\ \overset{\circ}{\mathrm{Grad}} & 0 \end{pmatrix}$ is further restricted to vector fields with vanishing divergence. This restriction is supposed to be a simplification. But is it in fact simpler than the non-vanishing divergence condition?

We recall from Definition 2.2.1 that $\iota_n(\mathrm{div})$ denotes the embedding from the null space of div into $L^2(\Omega)^3$. At first glance this suggests that we consider

$$\begin{pmatrix} 0 & \overline{\iota_n(\mathrm{div})^* \,\mathrm{Div}} \\ \overset{\circ}{\mathrm{Grad}}\, \iota_n(\mathrm{div}) & 0 \end{pmatrix}$$

in the spirit of the mother-and-descendant mechanism of Sect. A.1. This is, however, not feasible. We would want to identify the operator C as $\overset{\circ}{\mathrm{Grad}}\, \iota_n(\mathrm{div})$. Obviously, this operator is closed as a composition of a closed and a bounded operator. However, the answer to the following problem is a priori unclear:

Problem Is $\mathrm{dom}\left(\overset{\circ}{\mathrm{Grad}}\, \iota_n(\mathrm{div})\right)$ dense in $\ker(\mathrm{div}) = [\{0\}]\,\mathrm{div}$?

To by-pass the intricate issue of $\overset{\circ}{\mathrm{Grad}}\, \iota_n(\mathrm{div})$ being densely defined let alone the equality $\left(\overset{\circ}{\mathrm{Grad}}\, \iota_n(\mathrm{div})\right)^* = \overline{\iota_n(\mathrm{div})^* \,\mathrm{Div}}$, which is unclear as well, we introduce a slight variant of $\overset{\circ}{\mathrm{Grad}}\, \iota_n(\mathrm{div})$:

Proposition 2.3.7 *Let* $\Omega \subseteq \mathbb{R}^3$ *be open. Then the operator*[9]

$$\overset{\circ}{\mathrm{Grad}}_{\sigma,0} := \overset{\circ}{\mathrm{Grad}}\, \iota_n(\mathrm{div})|_{3\overset{\circ}{C}_\infty}$$

is a closable operator from $[\{0\}]\,\mathrm{div}$ *to* $L^2(\Omega)^{3\times3}$.

Proof This is clear, due to the closedness of $\overset{\circ}{\mathrm{Grad}}\, \iota_n(\mathrm{div})$. □

Having established closability of $\overset{\circ}{\mathrm{Grad}}_{\sigma,0}$, we need to confirm that this operator is densely defined as well. If we consider $[\{0\}]\,\mathrm{div}$ as the underlying Hilbert space, we run into the problem mentioned above. Thus, we are led to the following notion:

Definition 2.3.8 Let $\Omega \subseteq \mathbb{R}^3$ open. Then we define

$$L^2_\sigma(\Omega) := \overline{\mathrm{dom}\left(\iota_n(\mathrm{div})|_{\overset{\circ}{C}_\infty(\Omega)^3}\right)} = \overline{\overset{\circ}{C}_\infty(\Omega)^3 \cap [\{0\}]\,\mathrm{div}} \subseteq [\{0\}]\,\mathrm{div}.$$

---

[9]The subscript σ is a reminder of 'solenoidal', i.e., div-free.

With this space at hand, we may now define the appropriate operator for *any* open set Ω :

Theorem 2.3.9 *Let* $\Omega \subseteq \mathbb{R}^3$ *be open. Then* $\text{ran}(\overset{\circ}{\text{Grad}}_{\sigma,0}) \subseteq [\{0\}]$ trace *and the operator*

$$\overset{\circ}{\text{Grad}}_\sigma := \overline{\iota_n(\text{trace})^*\overset{\circ}{\text{Grad}}_{\sigma,0}|_{L^2_\sigma(\Omega)}}$$

is densely defined and closed.

Proof First we observe that for all $v \in \text{dom}\left(\overset{\circ}{\text{Grad}}_{\sigma,0}\right)$ we have

$$\text{trace}\,\overset{\circ}{\text{Grad}}_{\sigma,0}v = \text{div}\,v = 0.$$

This carries over to $v \in \text{dom}\left(\overline{\overset{\circ}{\text{Grad}}_{\sigma,0}}\right)$, where $\overset{\circ}{\text{Grad}}_{\sigma,0}$ is closable, by Proposition 2.3.7. Thus, the operator $\overset{\circ}{\text{Grad}}_\sigma$ is closed. The operator $\overset{\circ}{\text{Grad}}_\sigma$ is densely defined by definition of $L^2_\sigma(\Omega)$. $\qquad\square$

Theorem 2.3.9 asserts that

$$\overset{\circ}{\text{Grad}}_\sigma : \text{dom}(\overset{\circ}{\text{Grad}}_\sigma) \subseteq L^2_\sigma(\Omega) \to [\{0\}]\,\text{trace}$$

is densely defined and closed. Thus, $\overset{\circ}{\text{Grad}}_\sigma^*$ is densely defined and closed, as well. Moreover, by construction we obtain $\overset{\circ}{\text{Grad}}_\sigma \subseteq \iota_n(\text{trace})^*\text{Grad}\,\iota_n(\text{div})$ and so

$$\overset{\circ}{\text{Grad}}_\sigma^* \supseteq -\overline{\iota_n(\text{div})^*\,\text{Div}\,\iota_n(\text{trace})}.$$

Thus, in our framework the operator for the Stokes and the Oseen problem takes on the abstract form

$$\partial_0 M_0 + M_1 + \begin{pmatrix} 0 & \overset{\circ}{\text{Grad}}_\sigma^* \\ -\overset{\circ}{\text{Grad}}_\sigma & 0 \end{pmatrix}$$

with underlying Hilbert space $H := L^2_\sigma(\Omega) \oplus [\{0\}]$ trace and varying material laws described by continuous linear operators M_0, M_1 in H.

Example 2.3.10 We shall have a closer look at the (linear) Oseen system (see [17]), which is formally (disregarding any domain issues and boundary conditions) given by

$$\partial_0 \mu_* v + (v_0^\top \cdot \nabla)v - \text{Div}\,T + \text{grad}\,p = f,$$

$$T = \mathcal{C}\,\text{Grad}\,v,$$

complemented by

$$\operatorname{div} v = 0,$$

where v, T and p are to be determined, μ_*, $\mathcal{C}$ are operators to be specified later and f models a given external forcing term. The vector field v_0 is considered to be known, as well. We will rewrite the Oseen system starting with the observation

$$\operatorname{grad} p = \operatorname{Div} \operatorname{trace}^* p.$$

Thus, we arrive at

$$\partial_0 \mu_* v + (v_0^\top \cdot \nabla) v - \operatorname{Div} \left(T - \operatorname{trace}^* p \right) = f \qquad (2.3.11)$$

$$T = \mathcal{C} \operatorname{Grad} v$$

Next, to re-formulate the Oseen term $(v_0^\top \cdot \nabla) v$, we note

$$(v_0^\top \cdot \nabla) v = \left(\sum_{k=1}^{3} v_{0,k} \partial_k v_j \right)_j$$

$$= \left(\sum_{k=1}^{3} v_{0,k} \left(\partial_k v_j + \partial_j v_k \right) \right)_j - \left(\sum_{k=1}^{3} v_{0,k} \partial_j v_k \right)_j$$

$$= 2 v_0^\top \operatorname{Grad} v - \operatorname{grad} (v_0 \cdot v) + (\operatorname{grad} v_0)^\top v$$

$$= 2 v_0^\top \mathcal{C}^{-1} T - \operatorname{Div} \operatorname{trace}^* (v_0 \cdot v) + (\operatorname{grad} v_0)^\top v.$$

Using (2.3.11), we obtain

$$\partial_0 \mu_* v + (\operatorname{grad} v_0)^\top v + 2 v_0^\top \mathcal{C}^{-1} T - \operatorname{Div} \left(T - \operatorname{trace}^* (p - (v_0 \cdot v)) \right) = f,$$

$$T = \mathcal{C} \operatorname{Grad} v.$$

Furthermore, noting that

$$\left\langle w \,|\, \operatorname{Div} \operatorname{trace}^* (p - (v_0 \cdot v)) \right\rangle = \left\langle \overset{\circ}{\operatorname{Grad}} w \,|\, \operatorname{trace}^* (p - (v_0 \cdot v)) \right\rangle_{L^2(\Omega)^{3\times3}}$$

$$= \left\langle \operatorname{trace} \overset{\circ}{\operatorname{Grad}} w \,|\, (p - (v_0 \cdot v)) \right\rangle_{L^2(\Omega)}$$

$$= 0$$

for all $w \in \operatorname{dom}\left(\overset{\circ}{\operatorname{Grad}}_\sigma\right)$, the Oseen system reduces to

$$\partial_0 \mu_* v + (\operatorname{grad} v_0)^\top v + 2v_0^\top C^{-1} T + \operatorname{Grad}_\sigma^* T = f_0 \in L_\sigma^2(\Omega)$$

$$T = C \overset{\circ}{\operatorname{Grad}}_\sigma v$$

where f_0 is the orthogonal projection of f onto $L_\sigma^2(\Omega)$. Thus, we arrive at a proper evolutionary problem with

$$M_0 = \begin{pmatrix} \mu_* & 0 \\ 0 & 0 \end{pmatrix}, \quad M_1 = \begin{pmatrix} (\operatorname{grad} v_0)^\top & 2v_0^\top C^{-1} \\ 0 & C^{-1} \end{pmatrix}, \quad A = \begin{pmatrix} 0 & \operatorname{Grad}_\sigma^* \\ -\overset{\circ}{\operatorname{Grad}}_\sigma & 0 \end{pmatrix},$$

where with $v_0 = 0$ we recover the Stokes case. We remark here that for strictly positive definite C and μ_* in appropriate Hilbert spaces the conditions on M_0 and M_1 in (1.3.10) and (1.3.11) are satisfied, see also Proposition 2.1.5 for a similar case.

If we restrict C to the isotropic case, we have

$$C = c_0 \iota_n(\text{trace})^* \operatorname{sym} \iota_n(\text{trace}) = c_0 \operatorname{sym}_0,$$

where the parameter c_0 is in this context referred to as the viscosity. According to our abstract theory, see for instance Theorem 1.3.2, well-posedness of the system

$$\left(\partial_0 \begin{pmatrix} \mu_* & 0 \\ 0 & 0 \end{pmatrix} + \begin{pmatrix} (\operatorname{grad} v_0)^\top & 2v_0^\top C^{-1} \\ 0 & C^{-1} \end{pmatrix} + \begin{pmatrix} 0 & \operatorname{Grad}_\sigma^* \\ -\overset{\circ}{\operatorname{Grad}}_\sigma & 0 \end{pmatrix}\right) \begin{pmatrix} v \\ T \end{pmatrix} = \begin{pmatrix} f \\ 0 \end{pmatrix},$$

that is, continuous invertibility and causality of the (closure of the) operator

$$\partial_0 \begin{pmatrix} \mu_* & 0 \\ 0 & 0 \end{pmatrix} + \begin{pmatrix} (\operatorname{grad} v_0)^\top & 2v_0^\top C^{-1} \\ 0 & C^{-1} \end{pmatrix} + \begin{pmatrix} 0 & \operatorname{Grad}_\sigma^* \\ -\overset{\circ}{\operatorname{Grad}}_\sigma & 0 \end{pmatrix}$$

in $H_{\varrho,0}(\mathbb{R}; L_\sigma^2(\Omega) \oplus [\{0\}] \text{trace})$ for sufficiently large $\varrho > 0$ follows with $\mu_* : L_\sigma^2(\Omega) \to L_\sigma^2(\Omega)$ and $C : ([\{0\}] \text{trace}) \to ([\{0\}] \text{trace})$ continuous, selfadjoint, strictly positive definite, as well as $v_0 \in L^\infty(\Omega)^3$ such that $\operatorname{grad} v_0 \in L^\infty(\Omega)^{3\times3}$. The latter condition means that for every $\varphi \in \mathring{C}_\infty (\Omega)^3$ we have

$$\int_\Omega (\mathrm{div}\,\varphi)\,(x) \cdot v_0\,(x)\,dx = \int_\Omega \mathrm{trace}(\varphi\,(x)^\top \cdot z_0\,(x))\,dx \qquad (2.3.12)$$

for a (uniquely determined) $z_0 \in L^\infty\,(\Omega)^{3\times3}$. In this case,[10] $\mathrm{grad}\,v_0 := -z_0$.

We have used this example to demonstrate how much analytic effort can be required to repair a casually made "simplification" of the model, here from "small compressibility" to "incompressibility".

2.4   The Guyer–Krumhansl Model of Thermodynamics

In the previous two sections we considered cases where the evolutionary operator, that is, see (2.0.1) and (2.0.2),

$$\partial_0 M_0 + M_1 + \begin{pmatrix} 0 & -C^* \\ C & 0 \end{pmatrix} \qquad (2.4.1)$$

for some densely defined and closed linear operator C, involved C being a gradient acting on scalar or vector fields. By considering unitarily congruent versions or by using projection techniques, we derived new equations resulting in various partial differential equations. In this section, we present a kind of complementary way of looking at the spatial operator

$$A = \begin{pmatrix} 0 & -C^* \\ C & 0 \end{pmatrix}. \qquad (2.4.2)$$

In fact, a few models fall into a class we refer to as abstract grad-div systems, see [69]. These abstract grad-div systems are still of the standard type (2.4.1) with a spatial operator A of the form (2.4.2), but the closed, densely defined, linear operator C is itself composed

---

[10]In other words, the "distributional" gradient—as it is called—is representable by an element in $L^\infty\,(\Omega)^{3\times3}$. Condition (2.3.12) mimics the definition of $-\mathrm{grad}$ as the adjoint of $\mathring{\mathrm{div}}$ and extends it to include $L^\infty\,(\Omega)$-vector-fields. Note, however, that for Ω with bounded measure we have $L^\infty\,(\Omega) \subseteq L^2\,(\Omega)$. Thus only in the case of Ω with infinite measure do we need to properly re-define $\mathrm{grad}\,v_0$ for $v_0 \in L^\infty\,(\Omega)^3$.

of a list of operators $C_k : \mathrm{dom}\,(C) \subseteq H_0 \to H_k$, $k \in \{1, \ldots, n\}$, such that

$$C = \begin{pmatrix} C_1 \\ \vdots \\ C_n \end{pmatrix} : \mathrm{dom}\,(C) \subseteq H_0 \to H_1 \oplus \cdots \oplus H_n,$$

$$x \mapsto \begin{pmatrix} C_1 x \\ \vdots \\ C_n x \end{pmatrix},$$

where $H_0, H_1, \ldots, H_n$ are all Hilbert spaces. Whereas in the classical case of Sect. 2.1, we have $C_k = \partial_k$ for all $k \in \{1, \ldots, n\}$, in the present case of abstract grad-div systems the role of the partial derivatives is replaced by general (not necessarily closed or closable) linear operators. An example of this structure is given by a model of thermodynamics going back to Guyer and Krumhansl, which can be reformulated in our first order framework as follows.

2.4.1 The Spatial Operator of the Guyer–Krumhansl Model

For the original equations, we refer to e.g. [18–20].

The operator C to be considered in the following—assuming for example Dirichlet type boundary conditions—is

$$C = \begin{pmatrix} \mathring{\mathrm{div}} \\ -\mathrm{grad} \end{pmatrix} : \mathrm{dom}\,\left(\mathring{\mathrm{grad}}\right) \subseteq L^2\,(\Omega)^3 \to L^2\,(\Omega) \oplus L^2\,(\Omega)^{3\times 3}, \qquad (2.4.3)$$

$$v \mapsto \begin{pmatrix} \mathring{\mathrm{div}}\,v \\ -\mathring{\mathrm{grad}}\,v \end{pmatrix},$$

where the gradient is to be understood as the Jacobian. In this way, we get a peculiar interaction between systems discussed in Sects. 2.1 and 2.3.

Proposition 2.4.1 *The operator C given in (2.4.3) is densely defined and closed.*

Proof We recall from Proposition 2.3.2 that the operator $\mathrm{trp} = \frac{1}{3}\mathrm{trace}^*\mathrm{trace}$ is an orthogonal projector in $L^2\,(\Omega)^{3\times 3}$. Thus,

$$|\mathrm{trace}\,T|_{L^2(\Omega)} = \sqrt{3}\,|\mathrm{trp}\,T|_{L^2(\Omega)^{3\times 3}} \leq \sqrt{3}\,|T|_{L^2(\Omega)^{3\times 3}} \quad (T \in L^2(\Omega)^{3\times 3}),$$

so, with $T = \operatorname{grad} v$, noting that $(\operatorname{trace} \operatorname{grad}) v = \operatorname{div} v$, we obtain

$$|\operatorname{div} v|_{L^2(\Omega)} \leq \sqrt{3} |\operatorname{grad} v|_{L^2(\Omega)^{3 \times 3}} \quad (v \in \operatorname{dom}(\operatorname{grad})).$$

Specializing to $v \in \overset{\circ}{C}_\infty (\Omega)^3$ and using the definition of $\overset{\circ}{\operatorname{grad}}$ and div, we get

$$\left| \operatorname{div} v \right|_{L^2(\Omega)} \leq \sqrt{3} \left| \overset{\circ}{\operatorname{grad}} v \right|_{L^2(\Omega)^{3 \times 3}} \quad (v \in \operatorname{dom}(\overset{\circ}{\operatorname{grad}})). \tag{2.4.4}$$

Now, let $(v_k)_k$ be a sequence in $\operatorname{dom}(C)$ with $v_k \rightarrow v$ in $L^2(\Omega)^3$ and $C v_k \rightarrow w := (w_0, w_1)$ in $L^2(\Omega) \oplus L^2(\Omega)^{3 \times 3}$ as $k \rightarrow \infty$ for some $v \in L^2(\Omega)^3$, $w_0 \in L^2(\Omega)$ and $w_1 \in L^2(\Omega)^{3 \times 3}$. Then, by the closedness of $\overset{\circ}{\operatorname{grad}}$ we deduce that $v \in \operatorname{dom}(\overset{\circ}{\operatorname{grad}})$ and $-\overset{\circ}{\operatorname{grad}} v = w_1$. From (2.4.4) it follows that $(v_k)_k$ forms a Cauchy sequence in $\operatorname{dom}(\operatorname{div})$; by the closedness of div, there exists $\widetilde{w}_0 \in L^2(\Omega)$ with $v \in \operatorname{dom}(\operatorname{div})$ and $\operatorname{div} v = \widetilde{w}_0$. Since, $\operatorname{div} v_k \rightarrow w_0$ as $k \rightarrow \infty$, by hypothesis, we get $\widetilde{w}_0 = w_0$ and so $v \in \operatorname{dom}(C)$ with $Cv = w$ proving the closedness of C. $\square$

Thus we are led to consider evolutionary equations of the form

$$\left(\partial_0 M_0 + M_1 + \begin{pmatrix} 0 & -\begin{pmatrix} \operatorname{div} \\ -\overset{\circ}{\operatorname{grad}} \end{pmatrix}^* \\ \begin{pmatrix} \operatorname{div} \\ -\overset{\circ}{\operatorname{grad}} \end{pmatrix} & \begin{pmatrix} 0 & 0 \\ 0 & 0 \end{pmatrix} \end{pmatrix} \right) \begin{pmatrix} q \\ \theta \\ T \end{pmatrix} = \begin{pmatrix} f \\ g_1 \\ g_2 \end{pmatrix}, \tag{2.4.5}$$

for which our standard solution theory works as long as (1.3.1) is satisfied.

Before we address the Guyer–Krumhansl model of thermodynamics with its more specific material law, we consider the adjoint of C, that is, we seek a deeper understanding of

$$\begin{pmatrix} \operatorname{div} \\ -\overset{\circ}{\operatorname{grad}} \end{pmatrix}^*.$$

For this, we introduce the following notion:

Definition 2.4.2 Let $B \colon \operatorname{dom}(B) \subseteq H_0 \rightarrow H_1$ be densely defined and closed and define

$$\mathfrak{B} \colon \operatorname{dom}(B) \rightarrow H_1, \varphi \mapsto B\varphi,$$

which is a bounded operator from $\operatorname{dom}(B)$ (equipped with the graph norm of B) into H_1. We define

$$B^\diamond := \mathfrak{B}' R_{H_1}^{-1} \in \mathcal{B}(H_1, \operatorname{dom}(B)'),$$

where $\mathcal{B}' \colon H_1' \to \mathrm{dom}(B)'$ is the dual of $\mathcal{B}$ given by $\mathcal{B}' f \colon \mathrm{dom}(B) \ni \varphi \mapsto f(B\varphi)$ and $R_{H_1} \colon H_1' \to H_1$ is the Riesz isomorphism.

Note that if $\kappa \colon \mathrm{dom}(B) \hookrightarrow H_0$ denotes the canonical embedding, then $\kappa^\diamond = \kappa' R_{H_0}^{-1}$ embeds H_0 into $\mathrm{dom}(B)'$. Before we turn to a computation of the adjoint of C, we state a possible way of computing the adjoint of an (unbounded) operator by means of the notion just introduced (compare also Sect. B.5):

Proposition 2.4.3 *Let* $B \colon \mathrm{dom}(B) \subseteq H_0 \to H_1$ *be densely defined and closed and denote by* $\kappa \colon \mathrm{dom}(B) \to H_0$ *the canonical embedding. Then*

$$B^* = \{(\varphi, f) \in H_1 \oplus H_0; \, \kappa^\diamond f = B^\diamond \varphi\}.$$

In particular, if we identify $H_0 = H_0'$, *then* $\kappa'(v) = v$ *for all* $v \in H_0$ *and we have*

$$B^* = B^\diamond \cap (H_1 \oplus H_0).$$

Proof Let $(\varphi, f) \in H_1 \oplus H_0$. Then

$$B^* \varphi = f \iff \langle \varphi | Bv \rangle_{H_1} = \langle f | v \rangle_{H_0} \quad (v \in \mathrm{dom}(B))$$

$$\iff \left(R_{H_1}^{-1}\varphi\right)(Bv) = \langle f | v \rangle_{H_0} \quad (v \in \mathrm{dom}(B))$$

$$\iff \mathcal{B}' \left(R_{H_1}^{-1}\varphi\right)(v) = \langle f | \kappa(v) \rangle_{H_0} \quad (v \in \mathrm{dom}(B))$$

$$\iff \left(B^\diamond \varphi\right)(v) = \kappa'(R_{H_0}^{-1} f)(v) \quad (v \in \mathrm{dom}(B))$$

$$\iff \left(B^\diamond \varphi\right)(v) = \left(\kappa^\diamond f\right)(v) \quad (v \in \mathrm{dom}(B)).$$

Thus, the first formula for B^* is proved. Identifying H_0 with H_0' leads to $\kappa^\diamond = \kappa'$ and the second formula follows. $\qquad\square$

With the latter formula at hand, we see that for $C = \begin{pmatrix} \mathring{\mathrm{div}} \\ -\mathrm{grad} \end{pmatrix}$,

$$C^\diamond = \begin{pmatrix} \mathring{\mathrm{div}} \\ -\mathrm{grad} \end{pmatrix}^\diamond = \begin{pmatrix} \mathring{\mathrm{div}}^\diamond & -\mathring{\mathrm{grad}}^\diamond \end{pmatrix}$$

and so

$$C^* = \begin{pmatrix} \mathring{\mathrm{div}} \\ -\mathrm{grad} \end{pmatrix}^* \subseteq \begin{pmatrix} \mathring{\mathrm{div}}^\diamond & -\mathring{\mathrm{grad}}^\diamond \end{pmatrix}.$$

On the other hand, since, by definition, $\mathrm{grad} = -\mathring{\mathrm{div}}^*$, $\mathrm{div} = -\mathring{\mathrm{grad}}^*$ we also get

$$\left(-\mathrm{grad}\ \ \mathrm{div}\right) \subseteq \begin{pmatrix} \mathring{\mathrm{div}} \\ -\mathring{\mathrm{grad}} \end{pmatrix}^* \subseteq \left(\mathrm{div}^\diamond\ \ -\mathring{\mathrm{grad}}^\diamond\right).$$

In particular, $\mathrm{grad} \subseteq -\mathring{\mathrm{div}}^\diamond$ and $\mathrm{div} \subseteq -\mathring{\mathrm{grad}}^\diamond$. This motivates us to write grad and div again for their respective extensions $-\mathring{\mathrm{div}}^\diamond$ and $-\mathring{\mathrm{grad}}^\diamond$. With these conventions we have

$$-\begin{pmatrix} \mathring{\mathrm{div}} \\ -\mathring{\mathrm{grad}} \end{pmatrix}^* = \left(\mathrm{grad}\ \ -\mathrm{div}\right). \tag{2.4.6}$$

Thus, we arrive finally at

$$\left(\begin{matrix} 0 & -\begin{pmatrix} \mathring{\mathrm{div}} \\ -\mathring{\mathrm{grad}} \end{pmatrix}^* \\ \begin{pmatrix} \mathring{\mathrm{div}} \\ -\mathring{\mathrm{grad}} \end{pmatrix} & \begin{pmatrix} 0 & 0 \\ 0 & 0 \end{pmatrix} \end{matrix}\right) = \left(\begin{matrix} 0 & \left(\mathrm{grad}\ \ -\mathrm{div}\right) \\ \begin{pmatrix} \mathring{\mathrm{div}} \\ -\mathring{\mathrm{grad}} \end{pmatrix} & \begin{pmatrix} 0 & 0 \\ 0 & 0 \end{pmatrix} \end{matrix}\right). \tag{2.4.7}$$

2.4.2 The Guyer–Krumhansl Model

In this section, we will have a closer look at the Guyer–Krumhansl thermodynamical model for heat conduction. We consider (2.4.5) with the specific material laws

$$M_0 = \begin{pmatrix} \tau_0 \kappa^{-1} & \begin{pmatrix} 0 & 0 \end{pmatrix} \\ \begin{pmatrix} 0 \\ 0 \end{pmatrix} & \begin{pmatrix} \varrho_* c & 0 \\ 0 & 0 \end{pmatrix} \end{pmatrix}, \quad M_1 = \begin{pmatrix} \kappa^{-1} & \begin{pmatrix} 0 & 0 \end{pmatrix} \\ \begin{pmatrix} 0 \\ 0 \end{pmatrix} & \begin{pmatrix} 0 & 0 \\ 0 & C^{-1} \end{pmatrix} \end{pmatrix} \tag{2.4.8}$$

with

$$\mathcal{C} = \alpha_0 \mathrm{sym}_0 + \alpha_1\ \mathrm{trp} + \alpha_2 \mathrm{skew}, \tag{2.4.9}$$

where the parameters $\tau_0, \varrho_*, c, \alpha_0, \alpha_1, \alpha_2$ and κ are assumed to be positive numbers. Again the solution theory for this model in the sense of continuous invertibility of the (closure of the) operator given in (2.4.5) with (2.4.8) and (2.4.9) follows from our general solution theory (Theorem 1.3.2). Thus, we shall concentrate on formally reformulating this model as the more familiar second order formulation. Starting out with (2.4.5) and using (2.4.7)

$$
\left(
\partial_0
\begin{pmatrix}
\tau_0 \kappa^{-1} & \begin{pmatrix} 0 & 0 \end{pmatrix} \\
\begin{pmatrix} 0 \\ 0 \end{pmatrix} & \begin{pmatrix} \varrho_* c & 0 \\ 0 & 0 \end{pmatrix}
\end{pmatrix}
+
\begin{pmatrix}
\kappa^{-1} & \begin{pmatrix} 0 & 0 \end{pmatrix} \\
\begin{pmatrix} 0 \\ 0 \end{pmatrix} & \begin{pmatrix} 0 & 0 \\ 0 & C^{-1} \end{pmatrix}
\end{pmatrix}
\right.
$$

$$
\left.
+
\begin{pmatrix}
0 & \begin{pmatrix} \text{grad} & -\,\text{div} \end{pmatrix} \\
\begin{pmatrix} \overset{\circ}{\text{div}} \\ -\overset{\circ}{\text{grad}} \end{pmatrix} & \begin{pmatrix} 0 & 0 \\ 0 & 0 \end{pmatrix}
\end{pmatrix}
\right)
\begin{pmatrix} q \\ \theta \\ T \end{pmatrix}
=
\begin{pmatrix} f \\ g_1 \\ g_2 \end{pmatrix}
$$

and reading the system line by line, we get from the second block row a balance law

$$
\varrho_* c \partial_0 \theta + \overset{\circ}{\text{div}}\, q = g_1, \tag{2.4.10}
$$

along with

$$
T = C\left(\overset{\circ}{\text{grad}}\, q + g_2 \right). \tag{2.4.11}
$$

The first row gives

$$
\partial_0 \tau_0 q + q - \kappa\, \text{div}\, T + \kappa\, \text{grad}\, \theta = \kappa f. \tag{2.4.12}
$$

Let us now proceed to recover the original Guyer–Krumhansl model from the three equations (2.4.12), (2.4.10), (2.4.11). Since this requires unwanted additional regularity considerations, we do this only formally, since our contention is that the most appropriate model is already described by (2.4.5) and (2.4.8), (2.4.9).

The first aim is to consider (2.4.11) with $g_2 = 0$ and calculate

$$
\text{div}\, T = \text{div}\, C \overset{\circ}{\text{grad}}\, q.
$$

We state some formulas of combinatorical nature:

Proposition 2.4.4 *For all* $\varphi \in C_\infty(\Omega)^3$ *we have*

$$
\text{div}\,\text{sym}_0\, \text{grad}\, \varphi = \frac{1}{2}\Delta\varphi - \frac{1}{6} \text{grad}\,\text{div}\, \varphi,
$$

$$
\text{div}\,\text{trp}\, \text{grad}\, \varphi = \frac{1}{3}\,\text{grad}\,\text{div}\, \varphi,
$$

$$\text{div skew grad } \varphi = \frac{1}{2}\Delta\varphi - \frac{1}{2}\text{grad div } \varphi,$$

$$\text{div } \mathcal{C} \text{ grad } \varphi = \frac{\alpha_0 + \alpha_2}{2}\Delta + \frac{\alpha_0 + 2\alpha_1 - 3\alpha_2}{6}\text{ grad div } \varphi,$$

where $\mathcal{C}$ is given as in (2.4.9).

Proof We verify the formulas in turn. For the first one, we have

$$\text{div sym}_0 \text{ grad } \varphi = \text{div}\left(\frac{1}{2}\left(\partial_i\varphi_j + \partial_j\varphi_i - \frac{2}{3}\delta_{ij}\text{ div } \varphi\right)_{i,j\in\{1,2,3\}}\right)$$

$$= \left(\sum_{j=1}^{3}\partial_j\frac{1}{2}\left(\partial_i\varphi_j + \partial_j\varphi_i - \delta_{ij}\frac{2}{3}\text{ div } \varphi\right)\right)_{i\in\{1,2,3\}}$$

$$= \frac{1}{2}\Delta\varphi + \frac{1}{2}\text{grad div } \varphi - \frac{1}{3}\text{grad div } \varphi$$

$$= \frac{1}{2}\Delta\varphi + \frac{1}{6}\text{grad div } \varphi$$

Next, we compute

$$\text{div trp grad } \varphi = \text{div}\left(\frac{1}{3}\delta_{ij}\text{ div } \varphi\right)_{i,j\in\{1,2,3\}}$$

$$= \left(\sum_{j=1}^{3}\partial_j\frac{1}{3}\delta_{ij}\text{ div } \varphi\right)_{i\in\{1,2,3\}}$$

$$= \frac{1}{3}\text{grad div } \varphi.$$

Similarly, we calculate

$$\text{div skew grad } \varphi = \text{div skew}\left(\partial_k\varphi_j\right)_{j,k\in\{1,2,3\}}$$

$$= \frac{1}{2}\text{div}\left(\partial_k\varphi_j - \partial_j\varphi_k\right)_{j,k\in\{1,2,3\}}$$

$$= \frac{1}{2}\left(\sum_{k=1}^{3}\partial_k\left(\partial_k\varphi_j - \partial_j\varphi_k\right)\right)_{j\in\{1,2,3\}}$$

$$= \frac{1}{2}\left(\Delta - \text{grad div}\right)\varphi.$$

Thus, we get that

$$\operatorname{div} C \operatorname{grad} \varphi = \operatorname{div} \left(\alpha_0 \operatorname{sym}_0 + \alpha_1 \operatorname{trp} + \alpha_2 \operatorname{skew} \right) \operatorname{grad} \varphi$$

$$= \alpha_0 \operatorname{div} \operatorname{sym}_0 \operatorname{grad} \varphi + \alpha_1 \operatorname{div} \operatorname{trp} \operatorname{grad} \varphi +$$

$$+ \alpha_2 \operatorname{div} \operatorname{skew} \operatorname{grad} \varphi$$

$$= \alpha_0 \left(\frac{1}{2} \Delta \varphi + \frac{1}{6} \operatorname{grad} \operatorname{div} \varphi \right) + \alpha_1 \frac{1}{3} \operatorname{grad} \operatorname{div} \varphi +$$

$$+ \alpha_2 \frac{1}{2} (\Delta - \operatorname{grad} \operatorname{div}) \varphi$$

$$= \frac{\alpha_0 + \alpha_2}{2} \Delta \varphi + \frac{\alpha_0 + 2\alpha_1 - 3\alpha_2}{6} \operatorname{grad} \operatorname{div} \varphi. \qquad \Box$$

With these results at hand, we let

$$\mu_1 := \frac{\alpha_0 + \alpha_2}{2}, \quad \mu_2 := \frac{\alpha_0 + 2\alpha_1 - 3\alpha_2}{6},$$

and get from (2.4.12) with $f = 0$:

$$\partial_0 \tau_0 q + q - \kappa \mu_1 \Delta q - \kappa \mu_2 \operatorname{grad} \operatorname{div} q = -\kappa \operatorname{grad} \theta,$$

which is indeed the modified MCV model according to Guyer and Krumhansl, [18–20]. Note that the choices $\tau_0 = \mu_1 = \mu_2 = 0$ recover the standard Fourier law.

For well-posedness in terms of the new parameters μ_1 and μ_2, we recall that we need $\alpha_1, \alpha_2, \alpha_3 > 0$. Introducing $\lambda := \alpha_0 - 3\alpha_2 \in \mathbb{R}$, we get

$$\alpha_0 = \frac{3}{2} \left(\mu_1 + \frac{\lambda}{6} \right), \alpha_1 = 3 \left(\mu_2 - \frac{\lambda}{6} \right), \alpha_2 = \frac{1}{2} \left(\mu_1 - \frac{\lambda}{2} \right).$$

Thus, we need

$$6\mu_2 > \lambda, \quad 6\mu_1 > -\lambda, \quad \text{and} \quad 2\mu_1 > \lambda.$$

In particular we get

$$-6\mu_1 < \lambda < 2\mu_1$$

and

$$-6\mu_1 < \lambda < 6\mu_2$$

from which $\mu_1 > 0$ and $\frac{\alpha_0}{6} - \frac{\alpha_2}{2} = \lambda/6 \in]-\mu_1, \mu_2[$ follows. In any case, we need

$$\mu_2 > -\mu_1.$$

Remark 2.4.5 The degree of arbitrariness in the choice of α_0, α_2 (and depending upon this also α_1) can be utilized to restrict considerations to special cases such as $\alpha_2 = 0$, $\alpha_1 = 0$ or $\alpha_0 = 0$ via the mother-and-descendant mechanism, see Sect. A.1. This would lead, however, to further constraints on μ_1, μ_2. The case of $\alpha_2 = \alpha_1 = 0$ is intractable by the mother-and-descendant mechanism. It actually leads to a Stokes type system ($\text{div } q = \text{trace Grad } q = \text{trace grad } q = 0$), which requires a work-around[11] as discussed in Sect. 2.3.4.

2.5   The Equations of Electrodynamics

This section is devoted to Maxwell's equations, see [29, 30] (or [26, 62]). The classical equations are formulated in an open set $\Omega \subseteq \mathbb{R}^3$ and read

$$\left(\partial_0 M_0 + M_1 + \begin{pmatrix} 0 & -\text{curl} \\ \mathring{\text{curl}} & 0 \end{pmatrix} \right) \begin{pmatrix} E \\ H \end{pmatrix} = \begin{pmatrix} -J \\ K \end{pmatrix}$$

for suitable right-hand side J, K and M_0 and M_1 satisfying (1.3.1). We shall refer to this single equation as the Maxwell system. The operator $\mathring{\text{curl}}$ is the closure of the classical curl operator acting on smooth compactly supported vector-fields in $L^2(\Omega)^3$,

$$\text{curl}: \text{dom(curl)} \subseteq L^2(\Omega)^3 \to L^2(\Omega)^3$$

$$v \mapsto \begin{pmatrix} \partial_2 v_3 - \partial_3 v_2 \\ \partial_3 v_1 - \partial_1 v_3 \\ \partial_1 v_2 - \partial_2 v_1 \end{pmatrix},$$

where, for the classical curl operator

$$\text{dom(curl)} = \{v \in L^2(\Omega)^3; \partial_2 v_3 - \partial_3 v_2, \partial_3 v_1 - \partial_1 v_3, \partial_1 v_2 - \partial_2 v_1 \in L^2(\Omega)^3\}.$$

---

[11]Conversely, to by-pass the subtleties of this work-around for the Stokes or Oseen problem, the Guyer–Krumhansel model is used for numerical purposes under the heading of "artificial compressibility", see, for instance, [9] for an early reference.

Note that curl $=$ curl$\overset{\circ}{}$*. We shall further specify M_0 and M_1 in due course. At this point, we want to stress that the well-posedness of the classical Maxwell system—it is of the form (2.0.1) and (2.0.2)—follows from Theorem 1.3.2. Before we turn to a closer inspection of Maxwell system, we will identify it as being complementary to the equations of elasticity with symmetric stresses, see Sect. 2.3.3 in the sense that one has to single out the skew-symmetric part rather than the symmetric part as in the case of the equations of elasticity.

2.5.1   The Maxwell System as a Descendant of Elasticity

As in the case of symmetric stresses, we identify the standard Maxwell system as a descendant of (2.3.3). Thus, in this first section, we focus on A, the operator containing the spatial derivatives.

To formulate the forthcoming theorem, we need the following operator:

$$S: L^2(\Omega)^3 \ni \begin{pmatrix} \alpha_1 \\ \alpha_2 \\ \alpha_3 \end{pmatrix} \mapsto \frac{1}{\sqrt{2}} \begin{pmatrix} 0 & -\alpha_3 & \alpha_2 \\ \alpha_3 & 0 & -\alpha_1 \\ -\alpha_2 & \alpha_1 & 0 \end{pmatrix} \in \text{skew}[L^2(\Omega)^{3\times3}]. \qquad (2.5.1)$$

Lemma 2.5.1 *Let S be as in (2.5.1). Then S is unitary*[12] *and we have*

$$S^{-1} = S^* : \text{skew}[L^2(\Omega)^{3\times3}] \ni \begin{pmatrix} 0 & T_{12} & T_{13} \\ -T_{12} & 0 & T_{23} \\ -T_{13} & -T_{23} & 0 \end{pmatrix} \mapsto \sqrt{2} \begin{pmatrix} T_{32} \\ T_{13} \\ T_{21} \end{pmatrix} \in L^2(\Omega)^3.$$

Proof The assertion follows from the corresponding property for matrices, which can be confirmed by straightforward computation. □

Lemma 2.5.2 *Let H_0, H_1, H_2 be Hilbert spaces. Let $C: \text{dom}(C) \subseteq H_0 \to H_1$ be densely defined and closed, $B \in \mathcal{B}(H_1, H_2)$ and $\mathcal{D} \subseteq \text{dom}(C)$ a core for C. Moreover, we assume that BC is closable. Then $\mathcal{D}$ is a core for $\overline{BC}$.*

Proof Let $\varphi \in \text{dom}(\overline{BC})$, $\varepsilon > 0$. Then, by definition, there exists $\psi_1 \in \text{dom}(C)$ with the property

$$|\varphi - \psi_1|_{H_0} + |\overline{BC}\varphi - BC\psi_1|_{H_2} \leq \varepsilon.$$

---

[12]In the light of Footnote 8, $\frac{1}{\sqrt{2}}S^*T$ is—in the context of elasticity—referred to as the rotational or axial vector field associated with (the "rotational") T.

Since $\mathcal{D}$ is a core for C, there exists $\psi \in \mathcal{D}$ such that

$$|\psi_1 - \psi|_{H_0} + |C\psi_1 - C\psi|_{H_1} \leq \varepsilon.$$

Thus,

$$|\varphi - \psi| + |\overline{BC}\varphi - BC\psi| \leq |\varphi - \psi_1| + |\psi_1 - \psi| + |\overline{BC}\varphi - BC\psi_1| + \|B\| |C\psi_1 - C\psi|$$

$$\leq (1 + \max\{1, \|B\|\})\,\varepsilon,$$

which yields the assertion. $\qquad\qquad\qquad\qquad\qquad\qquad\qquad\qquad\qquad\qquad\qquad$ □

Next, we present a result that links the spatial operator of the Maxwell system to the skew-symmetric part of the spatial operator from (non-symmetric) elasticity:

Theorem 2.5.3 *With S from (2.5.1), we have the unitary congruence*

$$\frac{1}{\sqrt{2}} \begin{pmatrix} 0 & -\text{curl} \\ \overset{\circ}{\text{curl}} & 0 \end{pmatrix} = \begin{pmatrix} 1 & 0 \\ 0 & S^* \end{pmatrix} \begin{pmatrix} 0 & \text{div}\,\iota_{\text{skew}} \\ \iota_{\text{skew}}^* \overset{\circ}{\text{grad}} & 0 \end{pmatrix} \begin{pmatrix} 1 & 0 \\ 0 & S \end{pmatrix}.$$

Proof Since the operator

$$\begin{pmatrix} 1 & 0 \\ 0 & S \end{pmatrix} : L^2(\Omega)^3 \oplus L^2(\Omega)^3 \to L^2(\Omega)^3 \oplus \text{skew}[L^2(\Omega)^{3\times 3}]$$

is unitary (by Lemma 2.5.1) and since $(\text{div}\,\iota_{\text{skew}})^* = -\overline{\iota_{\text{skew}}^* \overset{\circ}{\text{grad}}}$ and $\text{curl}^* = \overset{\circ}{\text{curl}}$, it suffices to prove

$$\frac{1}{\sqrt{2}} \overset{\circ}{\text{curl}} = S^* \overline{\iota_{\text{skew}}^* \overset{\circ}{\text{grad}}}.$$

Since, $\overset{\circ}{C}_\infty(\Omega)^3$ is a core for both $\overset{\circ}{\text{curl}}$ (by definition) and $\overline{\iota_{\text{skew}}^* \overset{\circ}{\text{grad}}}$ (by Lemma 2.5.2), we need to establish

$$\frac{1}{\sqrt{2}} \overset{\circ}{\text{curl}}\, v = S^* \iota_{\text{skew}}^* \overset{\circ}{\text{grad}}\, v \quad \left(v \in \overset{\circ}{C}_\infty(\Omega)^3 \right).$$

For this, let $v \in \overset{\circ}{C}_\infty(\Omega)^3$ and compute with the help of Proposition 2.3.2:

$$\iota_{\text{skew}}^* \text{grad}\, v = \iota_{\text{skew}}^* \begin{pmatrix} \partial_1 v_1 & \partial_2 v_1 & \partial_3 v_1 \\ \partial_1 v_2 & \partial_2 v_2 & \partial_3 v_2 \\ \partial_1 v_3 & \partial_2 v_3 & \partial_3 v_3 \end{pmatrix}$$

$$= \frac{1}{2} \begin{pmatrix} 0 & \partial_2 v_1 - \partial_1 v_2 & \partial_3 v_1 - \partial_1 v_3 \\ \partial_1 v_2 - \partial_2 v_1 & 0 & \partial_3 v_2 - \partial_2 v_3 \\ \partial_1 v_3 - \partial_3 v_1 & \partial_2 v_3 - \partial_3 v_2 & 0 \end{pmatrix}.$$

Hence, from Lemma 2.5.1,

$$S^* \iota_{\text{skew}}^* \operatorname{grad} v = \frac{1}{\sqrt{2}} \begin{pmatrix} \partial_2 v_3 - \partial_3 v_2 \\ \partial_3 v_1 - \partial_1 v_3 \\ \partial_1 v_2 - \partial_2 v_1 \end{pmatrix} = \frac{1}{\sqrt{2}} \overset{\circ}{\operatorname{curl}} v. \qquad \square$$

Thus, we realize that "skew-symmetric elasticity", that is,

$$\left(\partial_0 \widetilde{M}_0 + \widetilde{M}_1 + \begin{pmatrix} 0 & \operatorname{div} \iota_{\text{skew}} \\ \iota_{\text{skew}}^* \overset{\circ}{\operatorname{grad}} & 0 \end{pmatrix} \right) \begin{pmatrix} \widetilde{E} \\ \widetilde{H} \end{pmatrix} = \begin{pmatrix} -\widetilde{J} \\ \widetilde{K} \end{pmatrix}$$

is up to a factor $\sqrt{2}$ unitarily congruent to the classical Maxwell system:

$$\left(\partial_0 M_0 + M_1 + \begin{pmatrix} 0 & -\operatorname{curl} \\ \overset{\circ}{\operatorname{curl}} & 0 \end{pmatrix} \right) \begin{pmatrix} E \\ H \end{pmatrix} = \begin{pmatrix} -J \\ K \end{pmatrix}, \qquad (2.5.2)$$

with appropriate M_0, M_1, E, H, J, K derived from the unitary congruence given by $\begin{pmatrix} 1 & 0 \\ 0 & S \end{pmatrix}$. The material law operators are commonly specialized to

$$M_0 = \begin{pmatrix} \varepsilon & 0 \\ 0 & \mu \end{pmatrix}, \quad M_1 = \begin{pmatrix} \sigma & 0 \\ 0 & 0 \end{pmatrix}. \qquad (2.5.3)$$

If M_0 and M_1 satisfy (1.3.1), we have of course well-posedness. For the case of classical materials as described by (2.5.3), requirement (1.3.1) amounts to μ being selfadjoint and strictly positive definite, ε being selfadjoint and, for some $c_0 > 0$, all $E \in L^2(\Omega)^3$ and all sufficiently large $\varrho \in \,]0, \infty[$,

$$\varrho \langle \varepsilon E | E \rangle_{L^2(\Omega)^3} + \langle \sigma E | E \rangle_{L^2(\Omega)^3} \geq c_0 \langle E | E \rangle_{L^2(\Omega)^3}.$$

Remark 2.5.4 The well-posedness for classical materials shows in particular that the divergence constraint

$$\operatorname{div} \varepsilon E = q, \quad \overset{\circ}{\operatorname{div}} \mu H = 0 \qquad (2.5.4)$$

with q a given charge density, which are commonly included in the system of Maxwell's equations, are superfluous. They can only amount to constraints on the data, which already provide a unique solution! We shall analyze this more deeply later.

If (1.3.1) is satisfied due to M_0 already being strictly positive definite, a sometimes useful reduction to $M_0 = 1$ can be achieved; for the definition of $\sqrt{M_0}$, we refer to Theorem B.8.2.

Proposition 2.5.5 *In* (2.5.2) *assume that M_0 is strictly positive definite. Then*

$$\left(\partial_0 M_0 + M_1 + \begin{pmatrix} 0 & -\operatorname{curl} \\ \overset{\circ}{\operatorname{curl}} & 0 \end{pmatrix}\right)$$

is similar to

$$\left(\partial_0 + \sqrt{M_0}^{-1} M_1 \sqrt{M_0}^{-1} + \sqrt{M_0}^{-1} \begin{pmatrix} 0 & -\operatorname{curl} \\ \overset{\circ}{\operatorname{curl}} & 0 \end{pmatrix} \sqrt{M_0}^{-1}\right).$$

In particular, for all $\begin{pmatrix} E \\ H \end{pmatrix}, \begin{pmatrix} -J \\ K \end{pmatrix} \in H_{\varrho,0}(\mathbb{R}; L^2(\Omega)^6)$ *and sufficiently large* $\varrho \in]0, \infty[$
so that (1.3.1) *is satisfied,*

$$\left(\partial_0 M_0 + M_1 + \begin{pmatrix} 0 & -\operatorname{curl} \\ \overset{\circ}{\operatorname{curl}} & 0 \end{pmatrix}\right) \begin{pmatrix} E \\ H \end{pmatrix} = \begin{pmatrix} -J \\ K \end{pmatrix}$$

if and only if $V := \sqrt{M_0} \begin{pmatrix} E \\ H \end{pmatrix}$ *satisfies*

$$\left(\partial_0 + \sqrt{M_0}^{-1} M_1 \sqrt{M_0}^{-1} + \sqrt{M_0}^{-1} \begin{pmatrix} 0 & -\operatorname{curl} \\ \operatorname{curl} & 0 \end{pmatrix} \sqrt{M_0}^{-1}\right) V$$

$$= \sqrt{M_0}^{-1} \begin{pmatrix} -J \\ K \end{pmatrix}. \qquad (2.5.5)$$

Proof It suffices to observe

$$\left(\partial_0 M_0 + M_1 + \begin{pmatrix} 0 & -\operatorname{curl} \\ \overset{\circ}{\operatorname{curl}} & 0 \end{pmatrix}\right)$$

$$= \sqrt{M_0} \Big(\partial_0 + \sqrt{M_0}^{-1} M_1 \sqrt{M_0}^{-1}$$

$$+ \sqrt{M_0}^{-1} \begin{pmatrix} 0 & -\operatorname{curl} \\ \overset{\circ}{\operatorname{curl}} & 0 \end{pmatrix} \sqrt{M_0}^{-1}\Big) \sqrt{M_0}. \qquad \square$$

Note here that $\sqrt{M_0}^{-1}\begin{pmatrix} 0 & -\operatorname{curl} \\ \mathring{\operatorname{curl}} & 0 \end{pmatrix}\sqrt{M_0}^{-1}$ is still skew-selfadjoint. In particular, for the case of classical materials, (2.5.3), we have

$$\sqrt{M_0}^{-1}\begin{pmatrix} 0 & -\operatorname{curl} \\ \mathring{\operatorname{curl}} & 0 \end{pmatrix}\sqrt{M_0}^{-1} = \begin{pmatrix} 0 & -\sqrt{\varepsilon}^{-1}\operatorname{curl}\sqrt{\mu}^{-1} \\ \sqrt{\mu}^{-1}\mathring{\operatorname{curl}}\sqrt{\varepsilon}^{-1} & 0 \end{pmatrix}$$

and

$$\sqrt{M_0}^{-1}M_1\sqrt{M_0}^{-1} = \begin{pmatrix} \sqrt{\varepsilon}^{-1}\sigma\sqrt{\varepsilon}^{-1} & 0 \\ 0 & 0 \end{pmatrix}.$$

Thus, we recover in this case the standard form (2.0.1), (2.0.2).

Remark 2.5.6 Following the abstract construction first applied in the acoustic case, see also Theorem 2.2.3, we may in this case also consider Maxwell's equations as evolutionary equations involving the skew-selfadjoint spatial operator

$$\begin{pmatrix} 0 & -\left|\sqrt{\mu}^{-1}\mathring{\operatorname{curl}}\sqrt{\varepsilon}^{-1}\right| \\ \left|\sqrt{\mu}^{-1}\mathring{\operatorname{curl}}\sqrt{\varepsilon}^{-1}\right| & 0 \end{pmatrix}$$

in

$$\overline{\sqrt{\varepsilon}^{-1}\operatorname{curl}\sqrt{\mu}^{-1}\left[L^2(\Omega)^3\right]} \oplus \overline{\sqrt{\varepsilon}^{-1}\operatorname{curl}\sqrt{\mu}^{-1}\left[L^2(\Omega)^3\right]}.$$

The corresponding Maxwell system operator acting in the complexification (see also Sect. 2.2.2) may be written simply as

$$\partial_0 + \mathrm{i}\left|\sqrt{\mu}^{-1}\mathring{\operatorname{curl}}\sqrt{\varepsilon}^{-1}\right|$$

or, assuming that the material law commutes with $\begin{pmatrix} 0 & -1 \\ 1 & 0 \end{pmatrix}$, we may even consider more generally

$$\partial_0 M_0 + M_1 + \mathrm{i}\left|\sqrt{\mu}^{-1}\mathring{\operatorname{curl}}\sqrt{\varepsilon}^{-1}\right|.$$

We mention this here only to point out how and under what assumptions Maxwell's equations can be properly simplified via a complex notation. We occasionally find in the case $M_0 = 1$, $M_1 = 0$, $\varepsilon = 1$, $\mu = 1$, the suggestion to write the original Maxwell's equations in a complex form

$$(\partial_0 + i\operatorname{curl})(E + iH) = -J + iK.$$

This works when $\Omega = \mathbb{R}^3$ since, in this case, $\mathring{\operatorname{curl}} = \operatorname{curl}$ or in other cases, where a boundary condition is chosen that makes curl selfadjoint, see [53, 54]. The general proper procedure, however, is as described above.

2.5.2 Non-classical Materials

In this section, we revisit the general situation (2.5.2). In addition to the case of classical materials, (2.5.3), more general situations are of interest, see [58, Section 4.1] as well as [74].

In the study of so-called electro-magnetic meta-materials we may have off-diagonal entries in M_0 and M_1 (such as bi-anisotropic media): If M_0 is non-block-diagonal, that is

$$M_0 = \begin{pmatrix} \varepsilon & \kappa \\ \kappa^* & \mu \end{pmatrix}$$

for $\kappa \neq 0$, one speaks of bi-anisotropic media. Particularly focussing on M_1, if

$$\frac{1}{2}(M_1 - M_1^*) = \begin{pmatrix} 0 & -\chi \\ \chi & 0 \end{pmatrix}$$

with χ selfadjoint, then one speaks of chiral media, and if

$$\frac{1}{2}(M_1 - M_1^*) = \begin{pmatrix} 0 & \tilde{\chi} \\ \tilde{\chi} & 0 \end{pmatrix}$$

with $\tilde{\chi}$ skew-selfadjoint then one speaks of Ω-media. In all these cases our solution theory (Theorem 1.3.2) yields well-posedness provided that (1.3.1) is satisfied.

In applications to more complex media, we also need to take higher order terms in the material law into account. For instance, the material law may be of the form

$$M\left(\partial_0^{-1}\right) = M_0 + \partial_0^{-1}M_1 + \partial_0^{-2}M_2\left(\partial_0^{-1}\right). \tag{2.5.6}$$

Here $M_2\left(\partial_0^{-1}\right)$ is a rational function of ∂_0^{-1} with operator-valued coefficients and we assume that $M_2\left(\partial_0^{-1}\right)$ is uniformly bounded for all sufficiently large $\varrho \in \,]0, \infty[$; such material laws are likely to occur after a so-called homogenization process, see [2, 89,

91, 92]. The material law taking the form (2.5.6) means that the resulting evolutionary equation—a perturbation as discussed in Theorem 1.3.8—is governed by the operator

$$\partial_0 M \left(\partial_0^{-1} \right) + A = \partial_0 M_0 + M_1 + \partial_0^{-1} M_2 \left(\partial_0^{-1} \right) + A.$$

General materials of the form (2.5.6) commonly result from a material law of rational form, in the simplest case of the form

$$p \left(\partial_0^{-1} \right) q \left(\partial_0^{-1} \right)^{-1},$$

where p and q are suitable polynomials. This form may for example stem from an ordinary differential equation model describing the material behavior. More generally, p and q are allowed to have operator (such as e.g. matrix-valued) coefficients so that in general

$$p \left(\partial_0^{-1} \right) q \left(\partial_0^{-1} \right)^{-1} \neq q \left(\partial_0^{-1} \right)^{-1} p \left(\partial_0^{-1} \right).$$

We need the following result.

Proposition 2.5.7 *Let H be a Hilbert space, $(a_k)_k$ a sequence in $\mathcal{B}(H)$, and $r > 0$. Assume that*

$$c_r := \sum_{k=0}^{\infty} \| a_k \| \, r^k < \infty.$$

Then, for all $\varrho \geq 1/r$,

$$p \left(\partial_{0,\varrho}^{-1} \right) := \sum_{k=0}^{\infty} a_k \left(\partial_{0,\varrho}^{-1} \right)^k$$

defines a bounded linear operator in $H_{\varrho,0}(\mathbb{R}; H)$. Moreover,

$$\left\| p \left(\partial_{0,\varrho}^{-1} \right) \right\|_{L\left(H_{\varrho,0}(\mathbb{R};H)\right)} \leq c_r.$$

Proof Let $f \in H_{\varrho,0}(\mathbb{R}; H)$. Then we compute with the help of Theorem 1.1.6 and Proposition 1.3.1:

$$\left\| p \left(\partial_{0,\varrho}^{-1} \right) f \right\|_{H_{\varrho,0}(\mathbb{R};H)} = \left\| \sum_{k=0}^{\infty} a_k \left(\partial_{0,\varrho}^{-1} \right)^k f \right\|_{H_{\varrho,0}(\mathbb{R};H)}$$

$$\leq \sum_{k=0}^{\infty} \|a_k\| \frac{1}{\varrho^k} \|f\|_{H_{\varrho,0}(\mathbb{R};H)}$$

$$\leq \sum_{k=0}^{\infty} \|a_k\| \, r^k \|f\|_{H_{\varrho,0}(\mathbb{R};H)} \leq c_r \|f\|_{H_{\varrho,0}(\mathbb{R};H)}. \qquad \square$$

Thus, in particular, let $p: \;]-r, r[\to \mathcal{B}(H)$ be analytic with the property that

$$\sum_{k=0}^{\infty} \left\| \frac{p^{(k)}(0)}{k!} \right\| r^k < \infty. \tag{2.5.7}$$

Then p defines a bounded linear operator in $H_{\varrho,0}(\mathbb{R}; H)$ via

$$p\left(\partial_0^{-1}\right) := \sum_{k=0}^{\infty} \frac{p^{(k)}(0)}{k!} \left(\partial_0^{-1}\right)^k \tag{2.5.8}$$

similar to Proposition 2.5.7.

Proposition 2.5.8 *Let H be a Hilbert space, $r > 0$, $p, q: \;]-r, r[\to \mathcal{B}(H)$ analytic with p satisfying (2.5.7), and $q(z)$ continuously invertible for all $z \in \;]-r, r[$. Then there exist $\varrho_0 > 0$ and an analytic $s: \;]-r', r'[\to \mathcal{B}(H)$ for some $r' \in \;]0, r[$ satisfying (2.5.7) with p and r replaced by s and r' respectively such that, for all $\varrho \geq \varrho_0$,*

$$p\left(\partial_0^{-1}\right) q \left(\partial_0^{-1}\right)^{-1}$$
$$= p(0)q(0)^{-1} + \partial_0^{-1}\left(p'(0)q(0)^{-1} - p(0)q(0)^{-1}q'(0)q(0)^{-1}\right) + \partial_0^{-2}s\left(\partial_0^{-1}\right).$$

In particular, if $M_0 := p(0)q(0)^{-1}$ and $M_1 := p'(0)q(0)^{-1} - p(0)q(0)^{-1}q'(0)q(0)^{-1}$ satisfy (1.3.1) and $A: \; \mathrm{dom}(A) \subseteq H \to H$ is skew-selfadjoint, then there exist $\varrho_0 > 0$ such that, for all $\varrho \geq \varrho_0$,

$$\overline{\partial_0 p\left(\partial_0^{-1}\right) q \left(\partial_0^{-1}\right)^{-1} + A}$$

is continuously invertible in $H_{\varrho,0}(\mathbb{R}; H)$ with causal inverse.

Proof By definition, we obtain (note that inversion is also an analytic function)

$$p\left(\partial_0^{-1}\right) = p(0) + \partial_0^{-1}p'(0) + \partial_0^{-2}s_p\left(\partial_0^{-1}\right),$$

$$q\left(\partial_0^{-1}\right)^{-1} = q(0)^{-1} - \partial_0^{-1}q(0)^{-1}q'(0)q(0)^{-1} + \partial_0^{-2}s_q\left(\partial_0^{-1}\right),$$

for some analytic s_p and s_q on $]-r', r'[$ for some $r' \in \,]0, r[$ with (2.5.7). Thus,

$$p\left(\partial_0^{-1}\right) q \left(\partial_0^{-1}\right)^{-1}$$

$$= p(0) q(0)^{-1} + \partial_0^{-1} \left(p'(0) q(0)^{-1} - p(0) q(0)^{-1} q'(0) q(0)^{-1}\right)$$

$$+ \partial_0^{-2} \left(s_{pq}\left(\partial_0^{-1}\right) + p\left(\partial_0^{-1}\right) s_q \left(\partial_0^{-1}\right) - p'(0) q(0)^{-1} q'(0) q(0)^{-1}\right).$$

The last statement follows from Theorem 1.3.8. □

Example 2.5.9 Returning to Maxwell's equations, a typical class of material laws of the form discussed in Proposition 2.5.8 is given by Drude–Lorentz type (see e.g. [74]) models, which are block diagonal but have higher order terms in place of ε:

$$M\left(\partial_0^{-1}\right) = \begin{pmatrix} \varepsilon_* \left(\partial_0^{-1}\right) & 0 \\ 0 & \mu \end{pmatrix}. \tag{2.5.9}$$

Allowing for several Drude–Lorentz type terms, we arrive at material laws of the form (2.5.9) with

$$\varepsilon_*\left(\partial_0^{-1}\right) = \varepsilon + \sum_{k=0}^{N} \left(\alpha_k \left(\kappa_k + \partial_0\right)^{-1} + \beta_k \left(\kappa_{k0} + \partial_0^2 - \kappa_{k1}\partial_0\right)^{-1}\right)$$

$$= \varepsilon + \partial_0^{-1} \sum_{k=0}^{N} \alpha_k \left(\kappa_k \partial_0^{-1} + 1\right)^{-1} +$$

$$+ \partial_0^{-2} \sum_{k=0}^{N} \beta_k \left(\left(\kappa_{k0}\partial_0^{-1} - \kappa_{k1}\right) \partial_0^{-1} + 1\right)^{-1} \tag{2.5.10}$$

where all coefficients κ_k, κ_{k0}, κ_{k1}, $k \in \{0, \ldots, N\}$, are bounded, linear operators in $L^2(\Omega)^3$, $N \in \mathbb{N}$. In classical Drude–Lorentz models these coefficients are just numbers. Noting that, by choosing $\varrho \in \,]0, \infty[$ sufficiently large, we have a convergent Neumann series expansion

$$\left(\beta_v \partial_0^{-1} + 1\right)^{-1} = \sum_{s=0}^{\infty} \left(\beta_v \partial_0^{-1}\right)^s$$

for any family $(\beta_\nu)_\nu$ of uniformly bounded linear mappings in $H_{\varrho,0}\left(\mathbb{R}, L^2\left(\Omega\right)^3\right)$, and we also see that the term

$$\sigma := \sum_{k=0}^{N} \alpha_k$$

acts as a conductivity term. The classical electromagnetic material (2.5.3) is recovered as a rather special case

$$\varepsilon_*\left(\partial_0^{-1}\right) = \varepsilon + \partial_0^{-1}\sigma.$$

Remark 2.5.10 In the Drude–Lorentz type model, consider the term

$$\beta_k\left(\kappa_{k0} + \partial_0^2 - \kappa_{k1}\partial_0\right)^{-1}. \tag{2.5.11}$$

Applying the so-called Fourier–Laplace transformation (see [32, 51] or [22, Corollary 2.5]), an integral transformation that yields that ∂_0 is unitarily equivalent to the multiplication operator $im + \varrho$, where m is the multiplication-by-argument-operator acting on $L^2(\mathbb{R})$, to (2.5.11) yields a function

$$z \mapsto \beta_k\left(\kappa_{k0} + z^2 - \kappa_{k1}z\right)^{-1}$$

which is small for all sufficiently large $\mathfrak{Re}\, z$, that is, for sufficiently large ϱ. This leads to the well-posedness of the evolutionary problem with this material law. Depending on the choice of parameters it may happen that (2.5.11) with ∂_0 replaced by $i\omega$ (i.e. $\varrho = 0$) becomes negative and dominant for a certain range of $\omega \in \mathbb{R}$ (frequency range). This is referred to as the negative dielectricity case (see e.g. [40]). If the permeability $\mu_*\left(\partial_0^{-1}\right)$ is also of Drude–Lorentz type and shares this property in the same frequency range, one speaks of double-negative media. The somewhat puzzling description as "negative dielectricity" (ε is never negative!) is due to the fact that the behavior for $\mathfrak{Re}\, z = 0$, that is, $\varrho = 0$, is used to label the media. This is a consequence of the fact that electrical engineers favor intuition derived from a non-physical limit case, the so-called time-harmonic waves,[13] for which $z = i\,\mathfrak{Im}\, z =: i\omega$, where ω is referred to as frequency.

---

[13]Time-harmonic waves only occur in the infinite future for an infinitely active time-harmonic source (limiting amplitude principle, see e.g. [34, 35]).

2.5.3   Some Decomposition Results

Having introduced all operators from vector analysis that are of interest, we shall elaborate a bit more on their interconnection, see also [48, 52] and [46] for other examples. In particular, we shall obtain more structural insights concerning the operator curl. We shall use these results to discuss the so-called extended Maxwell system in the subsequent section.

Proposition 2.5.11 *Let $\Omega \subseteq \mathbb{R}^3$ be open. Then*

$$
\begin{aligned}
\overset{\circ}{\mathrm{grad}} \left[\mathrm{dom}\left(\overset{\circ}{\mathrm{grad}}\,\right) \right] &\subseteq [\{0\}] \overset{\circ}{\mathrm{curl}}\,, \\
\mathrm{grad} \left[\mathrm{dom}\,(\mathrm{grad}) \right] &\subseteq [\{0\}] \mathrm{curl}, \\
\overset{\circ}{\mathrm{curl}} \left[\mathrm{dom}\left(\overset{\circ}{\mathrm{curl}}\,\right) \right] &\subseteq [\{0\}] \overset{\circ}{\mathrm{div}}\,, \\
\mathrm{curl}\,[\mathrm{dom}\,(\mathrm{curl})] &\subseteq [\{0\}] \mathrm{div}\,.
\end{aligned}
\tag{2.5.12}
$$

Proof Let $\varphi \in \overset{\circ}{C}_\infty(\Omega)$. Then, by Schwarz's lemma, we obtain

$$
\overset{\circ}{\mathrm{curl}}\,\overset{\circ}{\mathrm{grad}}\,\varphi = \overset{\circ}{\mathrm{curl}} \begin{pmatrix} \partial_1 \varphi \\ \partial_2 \varphi \\ \partial_3 \varphi \end{pmatrix}
$$

$$
= \begin{pmatrix} \partial_2 \partial_3 \varphi - \partial_3 \partial_2 \varphi \\ \partial_3 \partial_1 \varphi - \partial_1 \partial_3 \varphi \\ \partial_1 \partial_2 \varphi - \partial_2 \partial_1 \varphi \end{pmatrix} = 0.
$$

Similarly, for $v \in \overset{\circ}{C}_\infty(\Omega)^3$

$$
\overset{\circ}{\mathrm{div}}\,\overset{\circ}{\mathrm{curl}}\,v = \overset{\circ}{\mathrm{div}} \begin{pmatrix} \partial_2 v_3 - \partial_3 v_2 \\ \partial_3 v_1 - \partial_1 v_3 \\ \partial_1 v_2 - \partial_2 v_1 \end{pmatrix}
$$

$$
= \partial_1 \left(\partial_2 v_3 - \partial_3 v_2 \right) + \partial_2 \left(\partial_3 v_1 - \partial_1 v_3 \right) + \partial_3 \left(\partial_1 v_2 - \partial_2 v_1 \right)
$$

$$
= 0.
$$

Hence, let $\varphi \in \mathrm{dom}(\overset{\circ}{\mathrm{grad}}\,)$. Since $\overset{\circ}{\mathrm{grad}} := \overline{\mathrm{grad}\,|_{\overset{\circ}{C}_\infty(\Omega)}}$, there exists a sequence $(\varphi_n)_n$ in $\overset{\circ}{C}_\infty(\Omega)$ such that $\varphi_n \to \varphi$ and $\mathrm{grad}\,\varphi_n \to \overset{\circ}{\mathrm{grad}}\,\varphi$ as $n \to \infty$ in $L^2(\Omega)$ and $L^2(\Omega)^3$ respectively; for $n \in \mathbb{N}$ we have

$$
\overset{\circ}{\mathrm{curl}}\,\overset{\circ}{\mathrm{grad}}\,\varphi_n = 0.
$$

Since $\overset{\circ}{\operatorname{curl}}$ is closed, we deduce

$$\operatorname{grad}\varphi \in \operatorname{dom}(\overset{\circ}{\operatorname{curl}}), \quad \overset{\circ}{\operatorname{curl}}\operatorname{grad}\varphi = 0.$$

Hence,

$$\overset{\circ}{\operatorname{curl}}\operatorname{grad}\varphi = 0 \text{ for } \varphi \in \operatorname{dom}\left(\overset{\circ}{\operatorname{grad}}\right). \tag{2.5.13}$$

Similarly,

$$\operatorname{div}\overset{\circ}{\operatorname{curl}} v = 0 \text{ for } v \in \operatorname{dom}\left(\overset{\circ}{\operatorname{curl}}\right).$$

Thus, for $\varphi \in \operatorname{dom}(\operatorname{grad})$, we obtain for all $\Phi \in \operatorname{dom}(\overset{\circ}{\operatorname{curl}})$

$$\left\langle \overset{\circ}{\operatorname{curl}}\,\Phi\,\middle|\, \operatorname{grad}\varphi \right\rangle_{L^2(\Omega)^3} = -\left\langle \operatorname{div}\overset{\circ}{\operatorname{curl}}\,\Phi\,\middle|\,\varphi \right\rangle_{L^2(\Omega)}$$

$$= 0,$$

which implies that $\operatorname{grad}\varphi \in \operatorname{dom}(\overset{\circ}{\operatorname{curl}}{}^*) = \operatorname{dom}(\operatorname{curl})$ and

$$\operatorname{curl}\operatorname{grad}\varphi = 0.$$

Similarly, from (2.5.13) we obtain

$$\left\langle \operatorname{curl}\,\Phi\,\middle|\,\overset{\circ}{\operatorname{grad}}\varphi \right\rangle_{L^2(\Omega)^3} = 0$$

for all $\Phi \in \operatorname{dom}(\operatorname{curl})$, $\varphi \in \operatorname{dom}\left(\overset{\circ}{\operatorname{grad}}\right)$, from which it follows that

$$\operatorname{div}\operatorname{curl}\Phi = 0. \qquad\qquad \square$$

Next, we introduce

$$\mathcal{H}_D := \left\{ E \mid \operatorname{curl} E = 0,\ \operatorname{div} E = 0 \right\} = [\{0\}]\overset{\circ}{\operatorname{curl}} \cap [\{0\}]\operatorname{div},$$

$$\mathcal{H}_N := \left\{ H \mid \operatorname{curl} H = 0,\ \overset{\circ}{\operatorname{div}} H = 0 \right\} = [\{0\}]\operatorname{curl} \cap [\{0\}]\overset{\circ}{\operatorname{div}},$$

the spaces of Dirichlet fields (of electro-statics) and of Neumann fields (of magneto-statics), respectively. We speak of harmonic Dirichlet or Neumann fields since, due to the classical relation

$$\operatorname{curl} \operatorname{curl} - \operatorname{grad} \operatorname{div} = -\Delta,$$

such fields are in the null space of the (distributional) Laplacian Δ, that is, harmonic.

Corollary 2.5.12 *Let $\Omega \subseteq \mathbb{R}^3$ be open. Then we have the orthogonal decompositions*

$$[\{0\}] \operatorname{curl} = \overline{\operatorname{grad} \left[L^2 (\Omega) \right]} \oplus \mathcal{H}_N, \tag{2.5.14}$$

and

$$[\{0\}] \overset{\circ}{\operatorname{curl}} = \overline{\overset{\circ}{\operatorname{grad}} \left[L^2 (\Omega) \right]} \oplus \mathcal{H}_D. \tag{2.5.15}$$

Proof We only show the first equality as the second can be proven similarly. Since $L^2(\Omega)^3 = \overline{\operatorname{grad}[L^2(\Omega)]} \oplus [\{0\}]\overset{\circ}{\operatorname{div}}$ and $\overline{\operatorname{grad}[L^2(\Omega)]} \subseteq [\{0\}]\operatorname{curl}$, we obtain

$$[\{0\}] \operatorname{curl} = [\{0\}] \operatorname{curl} \cap \left(\overline{\operatorname{grad}[L^2(\Omega)]} \oplus [\{0\}]\overset{\circ}{\operatorname{div}} \right)$$

$$= \overline{\operatorname{grad}[L^2(\Omega)]} \oplus [\{0\}] \operatorname{curl} \cap [\{0\}]\overset{\circ}{\operatorname{div}}$$

$$= \overline{\operatorname{grad}[L^2(\Omega)]} \oplus \mathcal{H}_N,$$

which is the desired equality. $\qquad\square$

We arrive at the decomposition results needed in the next section:

Theorem 2.5.13 *Let $\Omega \subseteq \mathbb{R}^3$ open. Then we have the orthogonal decompositions*

$$\begin{aligned} L^2 (\Omega)^3 &= \overline{\operatorname{grad} \left[\operatorname{dom} (\operatorname{grad}) \right]} \oplus \mathcal{H}_N \oplus \overline{\overset{\circ}{\operatorname{curl}} \left[\operatorname{dom} \left(\overset{\circ}{\operatorname{curl}} \right) \right]}, \\ L^2 (\Omega)^3 &= \overline{\overset{\circ}{\operatorname{grad}} \left[\operatorname{dom} \left(\overset{\circ}{\operatorname{grad}} \right) \right]} \oplus \mathcal{H}_D \oplus \overline{\operatorname{curl} \left[\operatorname{dom} (\operatorname{curl}) \right]}. \end{aligned} \tag{2.5.16}$$

Moreover,

$$\begin{aligned} [\{0\}] \operatorname{div} &= \mathcal{H}_D \oplus \overline{\operatorname{curl} \left[\operatorname{dom} (\operatorname{curl}) \right]}, \\ [\{0\}] \overset{\circ}{\operatorname{div}} &= \mathcal{H}_N \oplus \overline{\overset{\circ}{\operatorname{curl}} \left[\operatorname{dom} \left(\overset{\circ}{\operatorname{curl}} \right) \right]}. \end{aligned} \tag{2.5.17}$$

Proof The result follows by substituting (2.5.14), (2.5.15) into the orthogonal decompositions

$$L^2 (\Omega)^3 = [\{0\}] \operatorname{curl} \oplus \overline{\overset{\circ}{\operatorname{curl}} \left[\operatorname{dom} \left(\overset{\circ}{\operatorname{curl}} \right) \right]},$$

$$L^2 (\Omega)^3 = [\{0\}] \overset{\circ}{\operatorname{curl}} \oplus \overline{\operatorname{curl} \left[\operatorname{dom} (\operatorname{curl}) \right]},$$

which are special cases of Theorem B.4.8. Then (2.5.17) again follows from Theorem B.4.8 by considering the special cases

$$L^2(\Omega)^3 = \overline{\operatorname{grad}\left[\operatorname{dom}\left(\mathring{\operatorname{grad}}\right)\right]} \oplus [\{0\}]\operatorname{div},$$

$$L^2(\Omega)^3 = \overline{\operatorname{grad}[\operatorname{dom}(\operatorname{grad})]} \oplus [\{0\}]\mathring{\operatorname{div}}.\qquad\square$$

2.5.4 The Extended Maxwell System

In this section we discuss a reformulation of the classical Maxwell's equations, which will result in the so-called extended Maxwell system. The extended Maxwell system is used to derive low-frequency asymptotics of Maxwell's equations, see [44, 45, 49, 50]. Moreover, it can be used to derive numerical strategies to solve Maxwell's equation, see [77]. Here, we will only consider the homogeneous, isotropic case.

The extended Maxwell system in the homogeneous isotropic case is (by re-scaling) of the form

$$(\partial_0 + A_{\text{eMax}}) \left(\begin{pmatrix} \varphi \\ E \\ \psi \\ H \end{pmatrix} \right) = \left(\begin{pmatrix} f_0 \\ f_1 \\ g_0 \\ g_1 \end{pmatrix} \right), \qquad (2.5.18)$$

where f_0, f_1, g_0, g_1 are given and

$$A_{\text{eMax}} := \begin{pmatrix} 0 & -W^* \\ W & 0 \end{pmatrix}$$

where

$$W := \begin{pmatrix} 0 & \operatorname{div} \\ -\operatorname{grad} & \mathring{\operatorname{curl}} \end{pmatrix},$$

that is, W is defined as

$$W: \operatorname{dom}(\operatorname{grad}) \oplus \left(\operatorname{dom}(\operatorname{div}) \cap \operatorname{dom}\left(\mathring{\operatorname{curl}}\right)\right)$$

$$\subseteq L^2(\Omega) \oplus L^2(\Omega)^3 \to L^2(\Omega) \oplus L^2(\Omega)^3$$

$$\begin{pmatrix} \varphi \\ E \end{pmatrix} \mapsto \begin{pmatrix} \operatorname{div} E \\ -\operatorname{grad}\varphi + \mathring{\operatorname{curl}} E \end{pmatrix}.$$

Clearly, (2.5.18) is an evolutionary equation with $M_0 = 1$ and $M_1 = 0$. Thus, our solution theory applies once we ascertain that W is densely defined, which is clear, and closed.

Lemma 2.5.14 *The operator W is closed.*

Proof Let $(\varphi_n)_n$ and $(E_n)_n$ be convergent in $L^2(\Omega)$ and $L^2(\Omega)^3$ to some φ and E, respectively, and assume that

$$(\operatorname{div} E_n)_n \quad \text{and} \quad \left(-\operatorname{grad} \varphi_n + \overset{\circ}{\operatorname{curl}} E_n\right)_n$$

converge to some $\psi \in L^2(\Omega)$ and $H \in L^2(\Omega)^3$, respectively. Then by the closedness of div, we deduce that $E \in \operatorname{dom}(\operatorname{div})$ and $\operatorname{div} E = \psi$. We define $\pi_r := \iota_r(\operatorname{grad})\iota_r(\operatorname{grad})^*$, the orthogonal projection onto $\overline{\operatorname{grad}[\operatorname{dom}(\operatorname{grad})]}$. Then, by Theorem 2.5.13, we obtain

$$-\operatorname{grad} \varphi_n = -\pi_r \operatorname{grad} \varphi_n = \pi_r \left(-\operatorname{grad} \varphi_n + \overset{\circ}{\operatorname{curl}} E_n\right) \to \pi_r H \quad (n \to \infty).$$

Hence, $\varphi \in \operatorname{dom}(\operatorname{grad})$ and $-\operatorname{grad} \varphi = \pi_r H$. Similarly, we obtain that

$$\overset{\circ}{\operatorname{curl}} E_n = (1 - \pi_r)\overset{\circ}{\operatorname{curl}} E_n = (1 - \pi_r)\left(-\operatorname{grad} \varphi_n + \overset{\circ}{\operatorname{curl}} E_n\right) \to (1 - \pi_r) H \quad (n \to \infty).$$

Thus, we have $(\varphi, E) \in \operatorname{dom}(\operatorname{grad}) \oplus \left(\operatorname{dom}(\operatorname{div}) \cap \operatorname{dom}\left(\overset{\circ}{\operatorname{curl}}\right)\right) = \operatorname{dom}(W)$ and

$$\operatorname{div} E = \psi, \quad -\operatorname{grad} \varphi + \overset{\circ}{\operatorname{curl}} E = H,$$

that is,

$$W\begin{pmatrix} \varphi \\ E \end{pmatrix} = \begin{pmatrix} \operatorname{div} E \\ -\operatorname{grad} \varphi + \overset{\circ}{\operatorname{curl}} E \end{pmatrix},$$

which proves the closedness of W. $\qquad \square$

We can provide some more insight into W^*:

Theorem 2.5.15

$$W^* = \begin{pmatrix} 0 & \operatorname{div} \\ -\operatorname{grad} & \overset{\circ}{\operatorname{curl}} \end{pmatrix}^* = \begin{pmatrix} 0 & \overset{\circ}{\operatorname{div}} \\ -\operatorname{grad} & \operatorname{curl} \end{pmatrix}.$$

Proof Let $\begin{pmatrix} \psi \\ H \end{pmatrix} \in \mathrm{dom}\,(W^*)$. Then we have (see Sect. B.5 or Proposition 2.4.3) for all $\begin{pmatrix} \varphi \\ E \end{pmatrix} \in \mathrm{dom}\,(W)$

$$\left\langle \begin{pmatrix} 0 & \mathrm{div} \\ -\,\mathrm{grad}\,\overset{\circ}{\mathrm{curl}} \end{pmatrix} \begin{pmatrix} \varphi \\ E \end{pmatrix} \middle| \begin{pmatrix} \psi \\ H \end{pmatrix} \right\rangle_{L^2(\Omega)\oplus L^2(\Omega)^3}$$

$$= \left\langle \begin{pmatrix} \varphi \\ E \end{pmatrix} \middle| \begin{pmatrix} 0 & \mathrm{div} \\ -\,\mathrm{grad}\,\overset{\circ}{\mathrm{curl}} \end{pmatrix}^* \begin{pmatrix} \psi \\ H \end{pmatrix} \right\rangle_{L^2(\Omega)\oplus L^2(\Omega)^3}$$

$$= \left\langle \begin{pmatrix} \varphi \\ E \end{pmatrix} \middle| \begin{pmatrix} 0 & \mathrm{div} \\ \mathrm{div}^\diamond & \mathrm{curl}^\diamond \end{pmatrix} \begin{pmatrix} \psi \\ H \end{pmatrix} \right\rangle_{L^2(\Omega)\oplus L^2(\Omega)^3}.$$

Specializing to $\varphi = 0$, we get

$$\left\langle \overset{\circ}{\mathrm{curl}}\,E \middle| H \right\rangle_{L^2(\Omega)^3} + \langle \mathrm{div}\,E | \psi \rangle_{L^2(\Omega)}$$

$$= \left\langle \begin{pmatrix} 0 & \mathrm{div} \\ -\,\mathrm{grad}\,\overset{\circ}{\mathrm{curl}} \end{pmatrix} \begin{pmatrix} 0 \\ E \end{pmatrix} \middle| \begin{pmatrix} \psi \\ H \end{pmatrix} \right\rangle_{L^2(\Omega)\oplus L^2(\Omega)^3}$$

$$= \left\langle \begin{pmatrix} 0 \\ E \end{pmatrix} \middle| \begin{pmatrix} 0 & \mathrm{div} \\ \mathrm{div}^\diamond & \overset{\circ}{\mathrm{curl}}^\diamond \end{pmatrix} \begin{pmatrix} \psi \\ H \end{pmatrix} \right\rangle_{L^2(\Omega)\oplus L^2(\Omega)^3} \qquad (2.5.19)$$

$$= \left\langle E \middle| \mathrm{div}^\diamond\,\psi + \overset{\circ}{\mathrm{curl}}{}^\diamond H \right\rangle_{L^2(\Omega)^3}.$$

Define $\pi_{\mathrm{r}} := \iota_{\mathrm{r}}(\mathrm{curl})\iota_{\mathrm{r}}(\mathrm{curl})^*$, the orthogonal projector onto $\overline{\mathrm{curl}\,[\mathrm{dom}\,(\mathrm{curl})]}$. Then, for all $E \in \mathrm{dom}(\overset{\circ}{\mathrm{curl}}) = \mathrm{dom}(\mathrm{curl}^*)$, we have (see Theorem B.4.8)

$$\overset{\circ}{\mathrm{curl}}\,E = \overset{\circ}{\mathrm{curl}}\,\pi_{\mathrm{r}}E$$

and (see Proposition 2.5.11)

$$\mathrm{div}\,\pi_{\mathrm{r}}E = 0.$$

In particular, if $E \in \mathrm{dom}(\overset{\circ}{\mathrm{curl}})$ then $\pi_{\mathrm{r}}E \in \mathrm{dom}(\mathrm{div}) \cap \mathrm{dom}(\overset{\circ}{\mathrm{curl}})$. Hence, from (2.5.19), for all $E \in \mathrm{dom}(\overset{\circ}{\mathrm{curl}})$

$$\langle \overset{\circ}{\mathrm{curl}}\,E | H \rangle_{L^2(\Omega)^3} = \langle \overset{\circ}{\mathrm{curl}}\,\pi_{\mathrm{r}}E | H \rangle_{L^2(\Omega)^3}$$

$$= \langle \overset{\circ}{\mathrm{curl}}\,\pi_{\mathrm{r}}E | H \rangle_{L^2(\Omega)^3} + \langle \mathrm{div}\,\pi_{\mathrm{r}}E | \psi \rangle_{L^2(\Omega)}$$

$$= \left\langle \pi_r E \,\middle|\, \mathrm{div}^\diamond\, \psi + \overset{\circ}{\mathrm{curl}}{}^\diamond H \right\rangle_{L^2(\Omega)^3}$$

$$= \left\langle E \,\middle|\, \pi_r \left(\mathrm{div}^\diamond\, \psi + \overset{\circ}{\mathrm{curl}}{}^\diamond H \right) \right\rangle_{L^2(\Omega)^3}.$$

Thus, $H \in \mathrm{dom}(\mathrm{curl})$ and

$$\mathrm{curl}\, H = \pi_r \left(\mathrm{div}^\diamond\, \psi + \overset{\circ}{\mathrm{curl}}{}^\diamond H \right).$$

Similarly, if $E \in \mathrm{dom}(\mathrm{div})$ then $(1 - \pi_r)\, E \in \mathrm{dom}(\overset{\circ}{\mathrm{curl}})$ with $\overset{\circ}{\mathrm{curl}}\,(1 - \pi_r)\, E = 0$ and $\mathrm{div}\,(1 - \pi_r)\, E = \mathrm{div}\, E$. Thus, from (2.5.19), for all $E \in \mathrm{dom}(\mathrm{div})$

$$\langle \mathrm{div}\, E \,|\, \psi \rangle_{L^2(\Omega)} = \langle \mathrm{div}\,(1 - \pi_r)\, E \,|\, \psi \rangle_{L^2(\Omega)}$$

$$= \langle \overset{\circ}{\mathrm{curl}}\,(1 - \pi_r)\, E \,|\, H \rangle_{L^2(\Omega)^3} + \langle \mathrm{div}\,(1 - \pi_r)\, E \,|\, \psi \rangle_{L^2(\Omega)}$$

$$= \left\langle (1 - \pi_r)\, E \,\middle|\, \mathrm{div}^\diamond\, \psi + \overset{\circ}{\mathrm{curl}}{}^\diamond H \right\rangle_{L^2(\Omega)^3}$$

$$= \left\langle E \,\middle|\, (1 - \pi_r) \left(\mathrm{div}^\diamond\, \psi + \overset{\circ}{\mathrm{curl}}{}^\diamond H \right) \right\rangle_{L^2(\Omega)^3}.$$

Thus, $\psi \in \mathrm{dom}\left(\overset{\circ}{\mathrm{grad}} \right)$ and

$$-\overset{\circ}{\mathrm{grad}}\, \psi = (1 - \pi_r) \left(\mathrm{div}^\diamond\, \psi + \overset{\circ}{\mathrm{curl}}{}^\diamond H \right).$$

Combining these two results we get

$$-\overset{\circ}{\mathrm{grad}}\, \psi + \mathrm{curl}\, H = \mathrm{div}^\diamond\, \psi + \overset{\circ}{\mathrm{curl}}{}^\diamond H.$$

Together we have

$$\begin{pmatrix} 0 & \mathrm{div} \\ -\overset{\circ}{\mathrm{grad}} & \mathrm{curl} \end{pmatrix} \begin{pmatrix} \psi \\ H \end{pmatrix} = \begin{pmatrix} 0 & \mathrm{div} \\ -\overset{\circ}{\mathrm{grad}} & \overset{\circ}{\mathrm{curl}} \end{pmatrix}^* \begin{pmatrix} \psi \\ H \end{pmatrix}.$$

So, $W^* \subseteq \begin{pmatrix} 0 & \mathrm{div} \\ -\overset{\circ}{\mathrm{grad}} & \mathrm{curl} \end{pmatrix}$. The converse inclusion is easy. □

As a consequence of Theorem 2.5.15 we have

$$
\begin{pmatrix} 0 & -W^* \\ W & 0 \end{pmatrix} = \begin{pmatrix} \begin{pmatrix} 0 & 0 \\ 0 & 0 \end{pmatrix} & \begin{pmatrix} 0 & -\overset{\circ}{\mathrm{div}} \\ \mathrm{grad} & -\mathrm{curl} \end{pmatrix} \\ \begin{pmatrix} 0 & \mathrm{div} \\ -\,\mathrm{grad} & \overset{\circ}{\mathrm{curl}} \end{pmatrix} & \begin{pmatrix} 0 & 0 \\ 0 & 0 \end{pmatrix} \end{pmatrix}
$$

$$
= \begin{pmatrix} 0 & 0 & 0 & -\overset{\circ}{\mathrm{div}} \\ 0 & 0 & \overset{\circ}{\mathrm{grad}} & -\mathrm{curl} \\ 0 & \mathrm{div} & 0 & 0 \\ -\,\mathrm{grad} & \overset{\circ}{\mathrm{curl}} & 0 & 0 \end{pmatrix}.
$$

Clearly,

$$
A_{\mathrm{eMax}} = A_{\mathrm{Max}} + A_{\mathrm{Ac}}.
$$

with the "Maxwell part"

$$
A_{\mathrm{Max}} := \begin{pmatrix} 0 & 0 & 0 & 0 \\ 0 & 0 & 0 & -\mathrm{curl} \\ 0 & 0 & 0 & 0 \\ 0 & \overset{\circ}{\mathrm{curl}} & 0 & 0 \end{pmatrix}
$$

and the "acoustic part"

$$
A_{\mathrm{Ac}} := \begin{pmatrix} 0 & 0 & 0 & -\overset{\circ}{\mathrm{div}} \\ 0 & 0 & \overset{\circ}{\mathrm{grad}} & 0 \\ 0 & \mathrm{div} & 0 & 0 \\ -\,\mathrm{grad} & 0 & 0 & 0 \end{pmatrix}.
$$

Due to their standard structure A_{Ac} and A_{Max} are both skew-selfadjoint. Moreover, it is a straightforward consequence of Proposition 2.5.11 that

$$
A_{\mathrm{Max}} A_{\mathrm{Ac}} = 0 \tag{2.5.20}
$$

and

$$
A_{\mathrm{Ac}} A_{\mathrm{Max}} = 0 \tag{2.5.21}
$$

on the domains $\mathrm{dom}\,(A_{\mathrm{Ac}}) = \mathrm{dom}\,(\mathrm{grad}) \oplus \mathrm{dom}\,(\mathrm{div}) \oplus \mathrm{dom}\,\left(\overset{\circ}{\mathrm{grad}}\right) \oplus \mathrm{dom}\,\left(\overset{\circ}{\mathrm{div}}\right)$ and $\mathrm{dom}\,(A_{\mathrm{Max}}) = L^2\,(\Omega) \oplus \mathrm{dom}\,\left(\overset{\circ}{\mathrm{curl}}\right) \oplus L^2\,(\Omega) \oplus \mathrm{dom}\,(\mathrm{curl})$, respectively.

It is possible to recover the original Maxwell's equations from the extended Maxwell system, see [65, Theorem 2.8] for the precise argument. To keep matters elementary and to avoid too much of a detour, we provide only a first step towards an equivalence of the extended Maxwell system and the original Maxwell's equations. For this, we use (2.5.21) to obtain:

Proposition 2.5.16 *Let $\varrho > 0$. Then for all $U \in \mathrm{dom}(\partial_0) \cap \mathrm{dom}(A_{\mathrm{eMax}})$*

$$(\partial_0 + A_{\mathrm{eMax}})\, U = (\partial_0 + A_{\mathrm{Ac}})\, (\partial_0 + A_{\mathrm{Max}})\, \partial_0^{-1} U.$$

Proof We compute

$$(\partial_0 + A_{\mathrm{Ac}})\, (\partial_0 + A_{\mathrm{Max}})\, \partial_0^{-1} U = \left(\partial_0^2 + A_{\mathrm{Ac}}\partial_0 + \partial_0 A_{\mathrm{Max}} + A_{\mathrm{Ac}}A_{\mathrm{Max}}\right)\partial_0^{-1} U$$

$$= (\partial_0 + A_{\mathrm{Ac}} + A_{\mathrm{Max}})\, U. \qquad \square$$

With this result, we can show the following.

Theorem 2.5.17 *Let $\varrho > 0$ and $f, g \in \mathrm{dom}(\partial_0)$ and $f \in \mathrm{dom}(\mathrm{div}), g \in \mathrm{dom}(\overset{\circ}{\mathrm{div}})$. Assume that $\varphi, \psi \in H_{\varrho,0}(\mathbb{R}; L^2(\Omega))$ and $E, H \in H_{\varrho,0}(\mathbb{R}; L^2(\Omega)^3)$ satisfy*

$$(\partial_0 + A_{\mathrm{eMax}})\left(\begin{pmatrix} \varphi \\ E \\ \psi \\ H \end{pmatrix}\right) = (\partial_0 + A_{\mathrm{Ac}})\left(\begin{pmatrix} 0 \\ f \\ 0 \\ g \end{pmatrix}\right).$$

Then

$$\left(\partial_0 + \begin{pmatrix} 0 & -\mathrm{curl} \\ \overset{\circ}{\mathrm{curl}} & 0 \end{pmatrix}\right)\begin{pmatrix} E \\ H \end{pmatrix} = \begin{pmatrix} \partial_0 f \\ \partial_0 g \end{pmatrix}.$$

Proof We put $F := (0, f, 0, g)$ and $U := (\varphi, E, \psi, H)$. Then, by Proposition 2.5.16, we deduce

$$(\partial_0 + A_{\mathrm{Ac}})\, F = (\partial_0 + A_{\mathrm{eMax}})\, U$$

$$= (\partial_0 + A_{\mathrm{Ac}})\, (\partial_0 + A_{\mathrm{Max}})\, \partial_0^{-1} U.$$

Hence, as A_{Ac} is skew-selfadjoint, we obtain

$$(\partial_0 + A_{Max}) \partial_0^{-1} U = F,$$

which leads to the desired equation.   □

Note that in [65], we also provided a relationship between the extended Maxwell system to the Dirac equation.

2.6   Coupled Physical Phenomena

Whereas, from a mathematical point of view, there is a relatively small number of equations modeling different physical phenomena, the multitude of coupled systems modeling the interaction of several physical phenomena is naturally much larger. However, there is a standard recipe[14] showing how to proceed in coupling different equations of evolutionary type.

2.6.1   The Coupling Recipe

We shall recall the general recipe as outlined in [58, Section 3.5]. In a first step, we just write down the different skew-selfadjoint spatial parts $(A_k : \mathrm{dom}\,(A_k) \subseteq H_k \to H_k)_{k \in \{0,\ldots,n\}}$ of some evolutionary equations in block diagonal form:

$$\mathcal{A} := \begin{pmatrix} A_0 & 0 & \cdots & 0 \\ 0 & \ddots & & \vdots \\ \vdots & & \ddots & 0 \\ 0 & \cdots & 0 & A_n \end{pmatrix}.$$

The operator $\mathcal{A} : \bigoplus_{k=0}^n \mathrm{dom}\,(A_k) \subseteq H \to H$ clearly inherits the skew-selfadjointness in $H := \bigoplus_{k \in \{0,\ldots,n\}} H_k$ from its skew-selfadjoint diagonal entries $A_k : \mathrm{dom}\,(A_k) \subseteq H_k \to H_k, k \in \{0, \ldots, n\}$. Combining this with simple material laws yields again an evolutionary equation:

$$(\partial_0 M_0 + M_1 + \mathcal{A}) U = F.$$

---

[14]In a sense, the construction of complex grad-div systems, such as the previously discussed Guyer–Krumhansel model (see Sect. 2.4), can also be considered as a coupling of different physical phenomena, which is, however, disjoint from the construction we discuss here.

With

$$U = \begin{pmatrix} U_0 \\ \vdots \\ \vdots \\ U_n \end{pmatrix}, \quad F = \begin{pmatrix} F_0 \\ \vdots \\ \vdots \\ F_n \end{pmatrix}, \quad M_j = \begin{pmatrix} M_{j,00} & 0 & \cdots & 0 \\ 0 & \ddots & & \vdots \\ \vdots & & \ddots & 0 \\ 0 & \cdots & 0 & M_{j,nn} \end{pmatrix}, \quad j \in \{0,1\},$$

we have merely combined the separate evolutionary problems

$$\left(\partial_0 M_{0,kk} + M_{1,kk} + A_k \right) U_k = F_k, \ k \in \{0, \ldots, n\},$$

into a single equation involving the block diagonal operator

$$\left(\partial_0 M_0 + M_1 + \mathcal{A} \right) = \begin{pmatrix} \partial_0 M_{0,00} + M_{1,00} + A_0 & 0 & \cdots & & 0 \\ 0 & \ddots & & & \vdots \\ \vdots & & \ddots & & 0 \\ 0 & & \cdots & 0 & \partial_0 M_{0,nn} + M_{1,nn} + A_n \end{pmatrix}.$$

Coupling occurs by allowing non-trivial off-diagonal entries in M_0 and M_1, so that

$$M_j := \begin{pmatrix} M_{j,00} & \cdots\cdots & M_{j,0n} \\ \vdots & \ddots & \vdots \\ \vdots & & \ddots & \vdots \\ M_{j,n0} & \cdots\cdots & M_{j,nn} \end{pmatrix}, \quad j \in \{0,1\}.$$

We shall illustrate this procedure by considering various particular cases. For other coupled phenomena and the corresponding analysis we refer to [37–39, 61, 67].

2.6.2 The Propagation of Cavities

In applications, the challenge is to show that the equations describing coupled physical phenomena can be written in the form outlined in Sect. 2.6.1. Here, we shall exemplify this by a system describing the propagation of cavities and related problems, see [43] for the precise formulation. First consider the usual model

$$\partial_0 \left(\partial_0^{-1} T \right) = T = C \overset{\circ}{\mathrm{Grad}}\, u + \beta \mathrm{trace}^* \varphi \tag{2.6.1}$$

$$\partial_0 \left(\partial_0^{-1} h \right) = h = \alpha \overset{\circ}{\mathrm{grad}}\, \varphi \tag{2.6.2}$$

$$\partial_0 \varrho_* \partial_0 u = \operatorname{Div} T + \varrho_* b \tag{2.6.3}$$

$$\varrho_* \kappa \partial_0^2 \varphi + \omega \partial_0 \varphi + \xi \varphi + \operatorname{trace} \beta^* \mathring{\operatorname{Grad}} u = \operatorname{div} h + \varrho_* \ell \tag{2.6.4}$$

in an open set $\Omega \subseteq \mathbb{R}^3$, where ℓ and b are given quantities and u, T, φ and h are the unknowns. Assuming strict positivity of C, ϱ_* and κ acting as selfadjoint operators in appropriate $L^2(\Omega)$-type spaces, we may reformulate the equations as follows. First, from (2.6.1)

$$\mathring{\operatorname{Grad}} u = C^{-1} T - C^{-1} \beta \operatorname{trace}^* \varphi \tag{2.6.5}$$

and substituting into (2.6.4) and applying ∂_0^{-1} we obtain

$$\varrho_* \kappa \partial_0 \varphi + \omega \varphi + \left(\xi - \operatorname{trace} \beta^* C^{-1} \beta \operatorname{trace}^* \right) \partial_0^{-1} \varphi$$

$$+ \operatorname{trace} \beta^* C^{-1} \partial_0^{-1} T = \operatorname{div} \partial_0^{-1} h + \varrho_* \partial_0^{-1} \ell. \tag{2.6.6}$$

Application of ∂_0^{-1} to (2.6.3) yields

$$\partial_0 \varrho_* u = \operatorname{Div} \partial_0^{-1} T + \varrho_* \partial_0^{-1} b. \tag{2.6.7}$$

The resulting system consisting of Eqs. (2.6.7), (2.6.5), (2.6.6), and (2.6.2) can be written as

$$
\left(\partial_0 \begin{pmatrix} \varrho_* & 0 & 0 & 0 \\ 0 & C^{-1} & 0 & 0 \\ 0 & 0 & \sqrt{\varrho_* \kappa} \sqrt{\varrho_*} & 0 \\ 0 & 0 & 0 & \alpha^{-1} \end{pmatrix} + \begin{pmatrix} 0 & 0 & 0 & 0 \\ 0 & 0 & -C^{-1}\beta \operatorname{trace}^* & 0 \\ 0 & \operatorname{trace}\beta^* C^{-1} & \omega & 0 \\ 0 & 0 & 0 & 0 \end{pmatrix} + \right.
$$

$$
+ \partial_0^{-1} \begin{pmatrix} 0 & 0 & 0 & 0 \\ 0 & 0 & 0 & 0 \\ 0 & 0 & \xi - \operatorname{trace}\beta^* C^{-1}\beta\operatorname{trace}^* & 0 \\ 0 & 0 & 0 & 0 \end{pmatrix} +
$$

$$
\left. + \begin{pmatrix} 0 & -\operatorname{Div} & 0 & 0 \\ -\mathring{\operatorname{Grad}} & 0 & 0 & 0 \\ 0 & 0 & 0 & \operatorname{div} \\ 0 & 0 & \mathring{\operatorname{grad}} & 0 \end{pmatrix} \right) \begin{pmatrix} u \\ \partial_0^{-1} T \\ \varphi \\ -\partial_0^{-1} h \end{pmatrix} = \begin{pmatrix} \varrho_* \partial_0^{-1} b \\ 0 \\ \varrho_* \partial_0^{-1} \ell \\ 0 \end{pmatrix}.
$$

This is clearly of the shape discussed in Sect. 2.6.1. In particular, the well-posedness and causality follows from Theorem 1.3.8 due to the presence of the operator term involving ∂_0^{-1}.

As a second example, we will discuss the following modification of the equations just discussed, where we highlighted the terms occurring additionally in red:

$$\partial_0 \left(\partial_0^{-1} T\right) = T = C\mathring{\mathrm{Grad}}\, u + D\mathrm{grad}\, \varphi + B\varphi \tag{2.6.8}$$

$$\partial_0 \left(\partial_0^{-1} h\right) = h = \alpha\mathrm{grad}\, \varphi + f\varphi + D^*\mathring{\mathrm{Grad}}\, u$$

$$\partial_0 \varrho_* u = \mathrm{Div} \left(\partial_0^{-1} T\right) + \varrho_* \partial_0^{-1} b \tag{2.6.9}$$

$$\varrho_* \kappa \partial_0^2 \varphi + \omega\partial_0\varphi + \xi\varphi + f^*\mathrm{grad}\, \varphi + B^*\mathring{\mathrm{Grad}}\, u = \mathrm{div}\, h + \varrho_* \ell, \tag{2.6.10}$$

where—as before—u, T, φ, and h are the unknowns and now C, α, D, B, ϱ_*, κ, ξ, and f are operators acting in appropriate $L^2(\Omega)$-spaces. We shall assume that

1. C, α, ϱ_*, and κ are selfadjoint,
2. α, C, ϱ_*, and $\left(C - D\alpha^{-1} D^*\right)$ are strictly positive definite.[15]

Then, we can reformulate Eqs. (2.6.8) for T and h as follows:

$$\mathring{\mathrm{Grad}}\, u + C^{-1} D\mathrm{grad}\, \varphi = C^{-1} T - C^{-1} B\, \varphi,$$

$$\alpha^{-1} D^*\mathring{\mathrm{Grad}}\, u + \mathrm{grad}\, \varphi = \alpha^{-1} h - \alpha^{-1} f\varphi,$$

or—in block operator matrix notation—

$$\begin{pmatrix} 1 & C^{-1}D \\ \alpha^{-1}D^* & 1 \end{pmatrix} \begin{pmatrix} \mathring{\mathrm{Grad}}\, u \\ \mathrm{grad}\, \varphi \end{pmatrix} = \begin{pmatrix} C^{-1} T - C^{-1} B\, \varphi \\ \alpha^{-1} h - \alpha^{-1} f\varphi \end{pmatrix},$$

which can be reduced to a triangular form

$$\begin{pmatrix} C^{-1}\left(C - D\alpha^{-1}D^*\right) & 0 \\ \alpha^{-1}D^* & 1 \end{pmatrix} \begin{pmatrix} \mathring{\mathrm{Grad}}\, u \\ \mathrm{grad}\, \varphi \end{pmatrix}$$

$$= \begin{pmatrix} C^{-1}T - C^{-1}B\, \varphi - C^{-1}D\left(\alpha^{-1}h - \alpha^{-1}f\varphi\right) \\ \alpha^{-1}h - \alpha^{-1}f\varphi \end{pmatrix}.$$

---

[15]The strict positive definiteness of $C - D\alpha^{-1}D^*$ can—perhaps more compactly—be expressed by the numerical range condition

$$w\left(C^{-1/2}D\alpha^{-1}D^*C^{-1/2}\right) \subseteq [0, 1[,$$

see also Definition B.6.1.

So, as $\left(C - D\alpha^{-1}D^*\right)$ is assumed to be strictly positive definite, we read off

$$\overset{\circ}{\mathrm{Grad}}\, u = \left(C - D\alpha^{-1}D^*\right)^{-1}\left(T - \left(B - D\alpha^{-1}f\right)\varphi - D\alpha^{-1}h\right) \qquad (2.6.11)$$

and

$$\overset{\circ}{\mathrm{grad}}\,\varphi = \alpha^{-1}h - \alpha^{-1}f\varphi - \alpha^{-1}D^*\,\overset{\circ}{\mathrm{Grad}}\, u. \qquad (2.6.12)$$

Substituting (2.6.11) into (2.6.12) we obtain

$$-\overset{\circ}{\mathrm{grad}}\,\varphi = \partial_0\left(\alpha^{-1} + \alpha^{-1}D^*\left(C - D\alpha^{-1}D^*\right)^{-1}D\alpha^{-1}\right)\left(-\partial_0^{-1}h\right) + \qquad (2.6.13)$$

$$+ \partial_0\alpha^{-1}D^*\left(C - D\alpha^{-1}D^*\right)^{-1}\left(\partial_0^{-1}T\right) +$$

$$+ \left(\alpha^{-1}f - \alpha^{-1}D^*\left(C - D\alpha^{-1}D^*\right)^{-1}\left(B - D\alpha^{-1}f\right)\right)\varphi.$$

Rewriting (2.6.11), we get

$$\partial_0\left(\left(C - D\alpha^{-1}D^*\right)^{-1}\left(\partial_0^{-1}T\right) + \left(C - D\alpha^{-1}D^*\right)^{-1}D\alpha^{-1}\left(-\partial_0^{-1}h\right)\right)$$

$$= \overset{\circ}{\mathrm{Grad}}\, u + \left(C - D\alpha^{-1}D^*\right)^{-1}\left(B - D\alpha^{-1}f\right)\varphi. \qquad (2.6.14)$$

Thus we have equations for the unknowns $-\partial_0^{-1}h$ and $\partial_0^{-1}T$. Next, we consider Eq. (2.6.10) to which we apply ∂_0^{-1} to obtain

$$\varrho_*\kappa\,\partial_0\varphi + \omega\varphi + \xi\,\partial_0^{-1}\varphi + f^*\partial_0^{-1}\overset{\circ}{\mathrm{grad}}\,\varphi$$

$$+ B^*\partial_0^{-1}\overset{\circ}{\mathrm{Grad}}\, u = -\,\mathrm{div}\left(-\partial_0^{-1}h\right) + \varrho_*\partial_0^{-1}\ell. \qquad (2.6.15)$$

Using (2.6.12) we get

$$f^*\partial_0^{-1}\overset{\circ}{\mathrm{grad}}\,\varphi = -f^*\alpha^{-1}\left(-\partial_0^{-1}h\right) - f^*\alpha^{-1}f\partial_0^{-1}\varphi - f^*\alpha^{-1}D^*\partial_0^{-1}\overset{\circ}{\mathrm{Grad}}\, u$$

and so from (2.6.15)

$$\varrho_*\kappa\,\partial_0\varphi + \omega\varphi + \xi\,\partial_0^{-1}\varphi - f^*\alpha^{-1}\left(-\partial_0^{-1}h\right) - f^*\alpha^{-1}f\partial_0^{-1}\varphi$$

$$+ \left(B^* - f^*\alpha^{-1}D^*\right)\partial_0^{-1}\overset{\circ}{\mathrm{Grad}}\, u = -\,\mathrm{div}\left(-\partial_0^{-1}h\right) + \varrho_*\partial_0^{-1}\ell. \qquad (2.6.16)$$

Now using (2.6.11) we obtain

$$
\left(B^* - f^* \alpha^{-1} D^* \right) \partial_0^{-1} \overset{\circ}{\mathrm{Grad}}\, u
$$

$$
= \left(B^* - f^* \alpha^{-1} D^* \right) \left(C - D\alpha^{-1} D^* \right)^{-1} \left(\partial_0^{-1} T \right)
$$

$$
- \left(B^* - f^* \alpha^{-1} D^* \right) \left(C - D\alpha^{-1} D^* \right)^{-1} \left(B - D\alpha^{-1} f \right) \partial_0^{-1} \varphi \qquad (2.6.17)
$$

$$
+ \left(B^* - f^* \alpha^{-1} D^* \right) \left(C - D\alpha^{-1} D^* \right)^{-1} D\alpha^{-1} \left(-\partial_0^{-1} h \right).
$$

Finally, using (2.6.16) and (2.6.17), we arrive at

$$
- \operatorname{div}\left(-\partial_0^{-1} h \right) + \varrho_* \partial_0^{-1} \ell = \varrho_* \kappa\, \partial_0 \varphi + \omega \varphi + \xi\, \partial_0^{-1} \varphi
$$

$$
- f^* \alpha^{-1} \left(-\partial_0^{-1} h \right) - f^* \alpha^{-1} f \partial_0^{-1} \varphi
$$

$$
+ \left(B^* - f^* \alpha^{-1} D^* \right) \partial_0^{-1} \overset{\circ}{\mathrm{Grad}}\, u
$$

$$
= \varrho_* \kappa\, \partial_0 \varphi + \omega \varphi
$$

$$
- f^* \alpha^{-1} \left(-\partial_0^{-1} h \right) + \left(\xi - f^* \alpha^{-1} f \right) \partial_0^{-1} \varphi
$$

$$
- \left(B^* - f^* \alpha^{-1} D^* \right) \left(C - D\alpha^{-1} D^* \right)^{-1} \left(B - D\alpha^{-1} f \right) \partial_0^{-1} \varphi
$$

$$
(2.6.18)
$$

$$
+ \left(B^* - f^* \alpha^{-1} D^* \right) \left(C - D\alpha^{-1} D^* \right)^{-1} \left(\partial_0^{-1} T \right)
$$

$$
+ \left(B^* - f^* \alpha^{-1} D^* \right) \left(C - D\alpha^{-1} D^* \right)^{-1} D\alpha^{-1} \left(-\partial_0^{-1} h \right).
$$

The resulting system consists of Eqs. (2.6.9), (2.6.14), (2.6.13), and (2.6.18) in the unknowns

$$
u, \partial_0^{-1} T, \varphi, \text{ and } - \partial_0^{-1} h.
$$

The equations are of the standard form

$$
\left(\partial_0 M_0 + M_1 + \partial_0^{-1} M_2 + \mathcal{A} \right)
\left(\begin{pmatrix} u \\ \partial_0^{-1} T \\ \varphi \\ -\partial_0^{-1} h \end{pmatrix} \right)
= \begin{pmatrix} F_0 \\ F_1 \end{pmatrix}.
\qquad (2.6.19)
$$

The spatial operator is

$$
\mathcal{A} = \begin{pmatrix} \begin{pmatrix} 0 & -\operatorname{Div} \\ -\operatorname{Grad} & 0 \end{pmatrix} & \begin{pmatrix} 0 & 0 \\ 0 & 0 \end{pmatrix} \\ \begin{pmatrix} 0 & 0 \\ 0 & 0 \end{pmatrix} & \begin{pmatrix} 0 & \operatorname{div} \\ \operatorname{grad} & 0 \end{pmatrix} \end{pmatrix}, \tag{2.6.20}
$$

and, collecting the coefficients in matrix form, we get (with $\kappa_* := \sqrt{\varrho_*}\kappa\sqrt{\varrho_*}$)

$$
M_0 = \begin{pmatrix} \begin{pmatrix} \varrho_* & 0 \\ 0 & \left(C - D\alpha^{-1}D^*\right)^{-1} \end{pmatrix} & M_{0,10}^* \\ M_{0,10} & \begin{pmatrix} \kappa_* & 0 \\ 0 & \alpha^{-1} + \alpha^{-1}D^*\left(C - D\alpha^{-1}D^*\right)^{-1}D\alpha^{-1} \end{pmatrix} \end{pmatrix}
$$

where

$$
M_{0,10} = \begin{pmatrix} 0 & 0 \\ 0 & \alpha^{-1}D^*\left(C - D\alpha^{-1}D^*\right)^{-1} \end{pmatrix}
$$

and

$$
M_1 = \begin{pmatrix} \begin{pmatrix} 0 & 0 \\ 0 & 0 \end{pmatrix} & -M_{1,10}^* \\ M_{1,10} & M_{1,11} \end{pmatrix}
$$

where

$$
M_{1,10} = \begin{pmatrix} 0 & \left(B^* - f^*\alpha^{-1}D^*\right)\left(C - D\alpha^{-1}D^*\right)^{-1} \\ 0 & 0 \end{pmatrix},
$$

$$
M_{1,11} = \begin{pmatrix} \omega & -M_{1,11,10}^* \\ M_{1,11,10} & 0 \end{pmatrix},
$$

where

$$
M_{1,11,10} = \alpha^{-1}f - \alpha^{-1}D^*\left(C - D\alpha^{-1}D^*\right)^{-1}\left(B - D\alpha^{-1}f\right).
$$

Finally, we have

$$M_2 = \begin{pmatrix} \begin{pmatrix} 0 & 0 \\ 0 & 0 \end{pmatrix} & \begin{pmatrix} 0 & 0 \\ 0 & 0 \end{pmatrix} \\ \begin{pmatrix} 0 & 0 \\ 0 & 0 \end{pmatrix} & M_{2,11} \end{pmatrix}$$

where

$$M_{2,11} = \begin{pmatrix} \xi - f^*\alpha^{-1}f - \left(B^* - f^*\alpha^{-1}D^*\right)\left(C - D\alpha^{-1}D^*\right)^{-1}\left(B - D\alpha^{-1}f\right) & 0 \\ 0 & 0 \end{pmatrix}.$$

Note that M_0 is congruent to

$$\begin{pmatrix} \begin{pmatrix} \varrho_* & 0 \\ 0 & \left(C - D\alpha^{-1}D^*\right)^{-1} \end{pmatrix} & \begin{pmatrix} 0 & 0 \\ 0 & 0 \end{pmatrix} \\ \begin{pmatrix} 0 & 0 \\ 0 & 0 \end{pmatrix} & \begin{pmatrix} \kappa_* & 0 \\ 0 & \alpha^{-1} \end{pmatrix} \end{pmatrix}$$

so that our assumptions on the operators involved clearly imply (1.3.10) and (1.3.11) (the M_2-part can be dealt with as a perturbation as in Theorem 1.3.8).

Highlighting positions with possibly non-zero entries we see the pattern

$$M_0 = \begin{pmatrix} \begin{pmatrix} \star & 0 \\ 0 & \star \end{pmatrix} & \begin{pmatrix} 0 & 0 \\ 0 & \star \end{pmatrix} \\ \begin{pmatrix} 0 & 0 \\ 0 & \star \end{pmatrix} & \begin{pmatrix} \star & 0 \\ 0 & \star \end{pmatrix} \end{pmatrix},$$

$$M_1 = \begin{pmatrix} \begin{pmatrix} 0 & 0 \\ 0 & 0 \end{pmatrix} & \begin{pmatrix} 0 & 0 \\ \star & 0 \end{pmatrix} \\ \begin{pmatrix} 0 & \star \\ 0 & 0 \end{pmatrix} & \begin{pmatrix} \star & \star \\ \star & 0 \end{pmatrix} \end{pmatrix},$$

$$\frac{1}{2}\left(M_1 + M_1^*\right) = \begin{pmatrix} \begin{pmatrix} 0 & 0 \\ 0 & 0 \end{pmatrix} & \begin{pmatrix} 0 & 0 \\ 0 & 0 \end{pmatrix} \\ \begin{pmatrix} 0 & 0 \\ 0 & 0 \end{pmatrix} & \begin{pmatrix} \star & 0 \\ 0 & 0 \end{pmatrix} \end{pmatrix},$$

$$M_2 = \left(\begin{pmatrix} \begin{pmatrix} 0 & 0 \\ 0 & 0 \end{pmatrix} & \begin{pmatrix} 0 & 0 \\ 0 & 0 \end{pmatrix} \\ \begin{pmatrix} 0 & 0 \\ 0 & 0 \end{pmatrix} & \begin{pmatrix} \bigstar & 0 \\ 0 & 0 \end{pmatrix} \end{pmatrix} \right).$$

The off-diagonal entries marked in red are those, through which the coupling between the "acoustic" and the elastic part occurs.

2.6.3   A Degenerate Reissner–Mindlin Plate Equation

As a similar system to the one encountered in the previous example, we next consider a degenerate Reissner–Mindlin plate equation. The equations are given for the same unknown $U = \left(u, \partial_0^{-1} T, \varphi, -\partial_0^{-1} h \right)$ as above with right-hand side $F = (0, 0, g, 0)$ in the usual form

$$(\partial_0 M_0 + M_1 + \mathcal{A}) U = F,$$

where

$$\mathcal{A} = \left(\begin{pmatrix} 0 & -\operatorname{Div} \\ -\overset{\circ}{\operatorname{Grad}} & 0 \end{pmatrix} \begin{pmatrix} 0 & 0 \\ 0 & 0 \end{pmatrix} \\ \begin{pmatrix} 0 & 0 \\ 0 & 0 \end{pmatrix} \begin{pmatrix} 0 & \operatorname{div} \\ \overset{\circ}{\operatorname{grad}} & 0 \end{pmatrix} \right)$$

and

$$M_0 := \left(\begin{pmatrix} 0 & 0 \\ 0 & C^{-1} \end{pmatrix} \begin{pmatrix} 0 & 0 \\ 0 & 0 \end{pmatrix} \\ \begin{pmatrix} 0 & 0 \\ 0 & 0 \end{pmatrix} \begin{pmatrix} \kappa_* & 0 \\ 0 & 0 \end{pmatrix} \right), \quad M_1 := \left(\begin{pmatrix} 0 & 0 \\ 0 & 0 \end{pmatrix} \begin{pmatrix} 0 & -1 \\ 0 & 0 \end{pmatrix} \\ \begin{pmatrix} 0 & 0 \\ 0 & 0 \end{pmatrix} \begin{pmatrix} \omega & 0 \\ 0 & 0 \end{pmatrix} \\ \begin{pmatrix} 1 & 0 \end{pmatrix} \end{pmatrix} \right).$$

We note here, that due to the zeros in the top left corner of both M_0 and M_1 our solution theory does not apply right away. Hence, the term "degenerate". We shall reformulate the equation to make our solution theory applicable.

Remark 2.6.1 We read off the first row equation:

$$\partial_0^{-1} h - \operatorname{Div} \partial_0^{-1} T = 0 \quad \text{or} \quad h = \operatorname{Div} T.$$

Similarly we get from the last equation

$$u + \overset{\circ}{\mathrm{grad}}\,\varphi = 0.$$

Eliminating u and h we get for the remaining two equations

$$\partial_0 C^{-1}\partial_0^{-1}T - \overset{\circ}{\mathrm{Grad}}\,u = \partial_0 C^{-1}\partial_0^{-1}T + \overset{\circ}{\mathrm{Grad}}\,\overset{\circ}{\mathrm{grad}}\,\varphi = 0$$

$$\partial_0 \kappa_* \varphi + \omega\varphi + \mathrm{div}\left(-\partial_0^{-1}h\right) = \partial_0 \kappa_* \varphi + \omega\varphi - \mathrm{div}\,\mathrm{Div}\,\partial_0^{-1}T = g.$$

Thus, we obtain

$$\left(\partial_0 \begin{pmatrix} \kappa_* & 0 \\ 0 & C^{-1} \end{pmatrix} + \begin{pmatrix} \omega & 0 \\ 0 & 0 \end{pmatrix} + \begin{pmatrix} 0 & -\mathrm{div}\,\mathrm{Div} \\ \overset{\circ}{\mathrm{Grad}}\,\overset{\circ}{\mathrm{grad}} & 0 \end{pmatrix}\right)\begin{pmatrix} \varphi \\ \partial_0^{-1}T \end{pmatrix} = \begin{pmatrix} g \\ 0 \end{pmatrix}.$$

Eliminating T, we arrive at the more familiar second order form

$$\kappa_* \partial_0^2 \varphi + \omega\partial_0 \varphi + \mathrm{div}\,\mathrm{Div}\,C\overset{\circ}{\mathrm{Grad}}\,\overset{\circ}{\mathrm{grad}}\,\varphi = \partial_0 g.$$

For simple isotropic homogeneous media we have

$$\mathrm{div}\,\mathrm{Div}\,C\overset{\circ}{\mathrm{Grad}}\,\overset{\circ}{\mathrm{grad}} \subseteq \Delta^2,$$

which results in the Kirchhoff plate equation, here in the Dirichlet case. Accepting this degenerate Reissner–Mindlin plate equation as a model comes, however, at a price (though not as substantial as for the Stokes problem, compare Sect. 2.3.4). Clearly,

$$\begin{pmatrix} 0 & -\mathrm{div}\,\mathrm{Div} \\ \overset{\circ}{\mathrm{Grad}}\,\overset{\circ}{\mathrm{grad}} & 0 \end{pmatrix} \text{ is skew-symmetric.}$$

However, in general $\overset{\circ}{\mathrm{Grad}}\,\overset{\circ}{\mathrm{grad}}$ is not closed and $\mathrm{div}\,\mathrm{Div}$ is in general neither closed nor equal to $\left(\overset{\circ}{\mathrm{Grad}}\,\overset{\circ}{\mathrm{grad}}\right)^*$. The usual work-around is to consider the closure $\overline{\overset{\circ}{\mathrm{Grad}}\,\overset{\circ}{\mathrm{grad}}}$ of the smaller operator $\overset{\circ}{\mathrm{Grad}}\,\mathrm{grad}\,|_{\overset{\circ}{C}_\infty(\Omega)} \subseteq \overset{\circ}{\mathrm{Grad}}\,\overset{\circ}{\mathrm{grad}}$ and replace $\mathrm{div}\,\mathrm{Div}$ by its extension $\overline{\overset{\circ}{\mathrm{Grad}}\,\overset{\circ}{\mathrm{grad}}}^*$.

Returning to the three-dimensional situation, we note that for general M_0, M_1 Eq. (2.6.19) is capable of describing various other models as special cases.

Example 2.6.2 We consider the Lord–Shulman model of thermo-elasticity (see [27] and [37, Section 3.2.1]), where now $\varphi = \partial_0^{-1}\theta$ with θ as temperature distribution and

$q = -T_0 h$ as the temperature flux, T_0 being a given real number. We shall not go into detail here, but merely observe the difference in the structure of the non-zero entries. In the Lord–Shulman case

$$
M_0 = \left(\begin{pmatrix} \begin{pmatrix} \star & 0 \\ 0 & \star \end{pmatrix} & \begin{pmatrix} 0 & 0 \\ \star & 0 \end{pmatrix} \\ \begin{pmatrix} 0 & \star \\ 0 & 0 \end{pmatrix} & \begin{pmatrix} \star & 0 \\ 0 & \star \end{pmatrix} \end{pmatrix} \right),
$$

whereas there is no coupling via

$$
M_1 = \left(\begin{pmatrix} \begin{pmatrix} 0 & 0 \\ 0 & 0 \end{pmatrix} & \begin{pmatrix} 0 & 0 \\ 0 & 0 \end{pmatrix} \\ \begin{pmatrix} 0 & 0 \\ 0 & 0 \end{pmatrix} & \begin{pmatrix} 0 & 0 \\ 0 & \star \end{pmatrix} \end{pmatrix} \right).
$$

The classical system of thermo-elasticity, [6], differs only by having the pattern

$$
M_0 = \left(\begin{pmatrix} \begin{pmatrix} \star & 0 \\ 0 & \star \end{pmatrix} & \begin{pmatrix} 0 & 0 \\ \star & 0 \end{pmatrix} \\ \begin{pmatrix} 0 & \star \\ 0 & 0 \end{pmatrix} & \begin{pmatrix} \star & 0 \\ 0 & 0 \end{pmatrix} \end{pmatrix} \right).
$$

This is indeed the same pattern as for the classical Biot system, see [5, 31], describing porous elastic media (again changing terminology and units with $\partial_0 \varphi$ as porosity!).

2.6.4 Thermo-Piezo-Electro-Magnetism

We conclude our journey on coupled physical phenomena with the equations describing the interconnected effects of heat propagation, elastic and electro-magnetic waves, see also [33] or [39] and the references therein. For this example we tag on the Maxwell block, which leads to

$$
\mathcal{A} := \left(\begin{pmatrix} \begin{pmatrix} 0 & -\operatorname{Div} \\ -\overset{\circ}{\operatorname{Grad}} & 0 \end{pmatrix} & \begin{pmatrix} 0 & 0 \\ 0 & 0 \end{pmatrix} \\ \begin{pmatrix} 0 & 0 \\ 0 & 0 \end{pmatrix} & \begin{pmatrix} 0 & \operatorname{div} \\ \overset{\circ}{\operatorname{grad}} & 0 \end{pmatrix} \\ \begin{pmatrix} \begin{pmatrix} 0 & 0 \\ 0 & 0 \end{pmatrix} & \begin{pmatrix} 0 & 0 \\ 0 & 0 \end{pmatrix} \end{pmatrix} \end{pmatrix} \quad \begin{pmatrix} \begin{pmatrix} 0 & 0 \\ 0 & 0 \end{pmatrix} \\ \begin{pmatrix} 0 & 0 \\ 0 & 0 \end{pmatrix} \\ \begin{pmatrix} 0 & -\operatorname{curl} \\ \overset{\circ}{\operatorname{curl}} & 0 \end{pmatrix} \end{pmatrix} \right).
$$

The material properties are described by

$$
M_0 := \left(\begin{array}{ccc}
\left(\begin{array}{cc}
\begin{pmatrix} \varrho_* & 0 \\ 0 & C^{-1} \end{pmatrix} & \begin{pmatrix} 0 & 0 \\ C^{-1}\lambda\Theta_0 & 0 \end{pmatrix} \\
\begin{pmatrix} 0 & \Theta_0\lambda^*C^{-1} \\ 0 & 0 \end{pmatrix} & \begin{pmatrix} \gamma_0 + \Theta_0\lambda^*C^{-1}\lambda\Theta_0 & 0 \\ 0 & \kappa_1 \end{pmatrix}
\end{array}\right) & M_{0,1}^* \\
M_{0,1} & \begin{pmatrix} \varepsilon + e^*C^{-1}e & 0 \\ 0 & \mu \end{pmatrix}
\end{array}\right),
$$

where

$$
M_{0,1} = \left(\begin{array}{cc}
\begin{pmatrix} 0 & e^*C^{-1} \\ 0 & 0 \end{pmatrix} & \begin{pmatrix} p\Theta_0 + e^*C^{-1}\lambda\Theta_0 & 0 \\ 0 & 0 \end{pmatrix}
\end{array}\right),
$$

and

$$
M_1 := \left(\begin{array}{cc}
\left(\begin{array}{cc}
\begin{pmatrix} \begin{pmatrix} 0 & 0 \\ 0 & 0 \end{pmatrix} & \begin{pmatrix} 0 & 0 \\ 0 & 0 \end{pmatrix} \\ \begin{pmatrix} 0 & 0 \\ 0 & 0 \end{pmatrix} & \begin{pmatrix} 0 & 0 \\ 0 & \kappa_0^{-1} \end{pmatrix} \\ \begin{pmatrix} 0 & 0 \\ 0 & 0 \end{pmatrix} & \begin{pmatrix} 0 & 0 \\ 0 & 0 \end{pmatrix} \end{pmatrix}
\end{array}\right) & \begin{pmatrix} \begin{pmatrix} 0 & 0 \\ 0 & 0 \end{pmatrix} \\ \begin{pmatrix} 0 & 0 \\ 0 & 0 \end{pmatrix} \\ \begin{pmatrix} \sigma & 0 \\ 0 & 0 \end{pmatrix} \end{pmatrix}
\end{array}\right).
$$

The pattern here is

$$
M_0 = \left(\begin{array}{cc}
\left(\begin{array}{cc}
\begin{pmatrix} \begin{pmatrix} \star & 0 \\ 0 & \star \end{pmatrix} & \begin{pmatrix} 0 & 0 \\ \star & 0 \end{pmatrix} \\ \begin{pmatrix} 0 & \star \\ 0 & 0 \end{pmatrix} & \begin{pmatrix} \star & 0 \\ 0 & \star \end{pmatrix} \\ \begin{pmatrix} 0 & \star \\ 0 & 0 \end{pmatrix} & \begin{pmatrix} \star & 0 \\ 0 & 0 \end{pmatrix} \end{pmatrix}
\end{array}\right) & \begin{pmatrix} \begin{pmatrix} 0 & 0 \\ \star & 0 \end{pmatrix} \\ \begin{pmatrix} \star & 0 \\ 0 & 0 \end{pmatrix} \\ \begin{pmatrix} \star & 0 \\ 0 & \star \end{pmatrix} \end{pmatrix}
\end{array}\right),
$$

whereas there is no coupling via

$$
M_1 = M_1^* = \left(\begin{array}{cc}
\left(\begin{array}{cc}
\begin{pmatrix} \begin{pmatrix} 0 & 0 \\ 0 & 0 \end{pmatrix} & \begin{pmatrix} 0 & 0 \\ 0 & 0 \end{pmatrix} \\ \begin{pmatrix} 0 & 0 \\ 0 & 0 \end{pmatrix} & \begin{pmatrix} 0 & 0 \\ 0 & \star \end{pmatrix} \\ \begin{pmatrix} 0 & 0 \\ 0 & 0 \end{pmatrix} & \begin{pmatrix} 0 & 0 \\ 0 & 0 \end{pmatrix} \end{pmatrix}
\end{array}\right) & \begin{pmatrix} \begin{pmatrix} 0 & 0 \\ 0 & 0 \end{pmatrix} \\ \begin{pmatrix} 0 & 0 \\ 0 & 0 \end{pmatrix} \\ \begin{pmatrix} \star & 0 \\ 0 & 0 \end{pmatrix} \end{pmatrix}
\end{array}\right).
$$

We shall verify the solvability condition for these operators M_0 and M_1.

Theorem 2.6.3 *Assume that* $\varrho_*, \varepsilon, \mu, C, \gamma_0, \kappa_1, \Theta_0$ *are selfadjoint with* ε, κ_1 *non-negative and* $\varrho_*, \mu, C, \gamma_0$ *strictly positive definite together with*

$$\varrho\left(\varepsilon - \Theta_0 p^* \gamma_0^{-1} p \Theta_0\right) + \sigma, \varrho\kappa_1 + \kappa_0^{-1} \text{ strictly positive definite}$$

uniformly for all sufficiently large $\varrho \in]0, \infty[$ *. Then,* M_0 *and* M_1 *satisfy the condition (1.3.1) and hence, the corresponding problem of thermo-piezo-electricity is a well-posed evolutionary problem in the sense of Theorem 1.3.2.*

Proof Obviously, M_0 is selfadjoint. Moreover, since ϱ_* and μ are strictly positive definite and $\varrho\kappa_1 + \kappa_0^{-1}$ strictly positive definite for all sufficiently large ϱ, the only thing which is left to show is that

$$\varrho\begin{pmatrix} C^{-1} & C^{-1}e & C^{-1}\lambda\Theta_0 \\ e^*C^{-1} & \varepsilon + e^*C^{-1}e & p\Theta_0 + e^*C^{-1}\lambda\Theta_0 \\ \Theta_0\lambda^*C^{-1} & \Theta_0 p^* + \Theta_0\lambda^*C^{-1}e & \gamma_0 + \Theta_0\lambda^*C^{-1}\lambda\Theta_0 \end{pmatrix} + \begin{pmatrix} 0 & 0 & 0 \\ 0 & \sigma & 0 \\ 0 & 0 & 0 \end{pmatrix}$$

strictly positive definite   (2.6.21)

for all sufficiently large ϱ. In other words, in the common block structure of M_0 and M_1, where non-zero entries are highlighted with $\cdot$, we only focus on the entries $\odot$:

$$\begin{pmatrix} \begin{pmatrix} \begin{pmatrix} \cdot & 0 \\ 0 & \odot \end{pmatrix} & \begin{pmatrix} 0 & 0 \\ \odot & 0 \end{pmatrix} \end{pmatrix} & \begin{pmatrix} \begin{pmatrix} 0 & 0 \\ \odot & 0 \end{pmatrix} \end{pmatrix} \\ \begin{pmatrix} \begin{pmatrix} 0 & \odot \\ 0 & 0 \end{pmatrix} & \begin{pmatrix} \odot & 0 \\ 0 & \cdot \end{pmatrix} \end{pmatrix} & \begin{pmatrix} \begin{pmatrix} \odot & 0 \\ 0 & 0 \end{pmatrix} \end{pmatrix} \\ \begin{pmatrix} \begin{pmatrix} 0 & \odot \\ 0 & 0 \end{pmatrix} & \begin{pmatrix} \odot & 0 \\ 0 & 0 \end{pmatrix} \end{pmatrix} & \begin{pmatrix} \begin{pmatrix} \odot & 0 \\ 0 & \cdot \end{pmatrix} \end{pmatrix} \end{pmatrix}.$$

By symmetric Gauss steps (eliminating off diagonal entries) as a congruence transformation we get that the operator in (2.6.21) is congruent to

$$\varrho\begin{pmatrix} C^{-1} & 0 & 0 \\ 0 & \varepsilon & p\Theta_0 \\ 0 & \Theta_0 p^* & \gamma_0 \end{pmatrix} + \begin{pmatrix} 0 & 0 & 0 \\ 0 & \sigma & 0 \\ 0 & 0 & 0 \end{pmatrix},$$

which itself is congruent by another symmetric Gauss step to

$$\varrho \begin{pmatrix} C^{-1} & 0 & 0 \\ 0 & \varepsilon - \Theta_0 p^* \gamma_0^{-1} p \Theta_0 & 0 \\ 0 & 0 & \gamma_0 \end{pmatrix} + \begin{pmatrix} 0 & 0 & 0 \\ 0 & \sigma & 0 \\ 0 & 0 & 0 \end{pmatrix}.$$

The latter operator is then strictly positive definite by assumption and so the assertion follows. □

Remark 2.6.4

(1) Note that, due to the generality of the assumptions, limit cases such as $\varepsilon = \Theta_0 p^* \gamma_0^{-1} p \Theta_0$ and σ strictly positive definite (eddy current case) are also covered by the theorem.
(2) To include piezo-magnetic effects, we would need to modify M_0 to allow for a pattern of the form

$$M_0 = \begin{pmatrix} \left(\begin{pmatrix} \star & 0 \\ 0 & \star \end{pmatrix} \begin{pmatrix} 0 & 0 \\ \star & 0 \end{pmatrix} \right) & \left(\begin{pmatrix} 0 & \star \\ \star & 0 \end{pmatrix} \right) \\ \left(\begin{pmatrix} 0 & \star \\ 0 & 0 \end{pmatrix} \begin{pmatrix} \star & 0 \\ 0 & \star \end{pmatrix} \right) & \left(\begin{pmatrix} \star & 0 \\ 0 & 0 \end{pmatrix} \right) \\ \left(\begin{pmatrix} 0 & \star \\ \star & 0 \end{pmatrix} \begin{pmatrix} \star & 0 \\ 0 & 0 \end{pmatrix} \right) & \left(\begin{pmatrix} \star & 0 \\ 0 & \star \end{pmatrix} \right) \end{pmatrix}.$$

If (1.3.1) is satisfied, we get a well-posedness result for the full thermo-piezo-electromagnetism.

But What About the Main Stream?

3

In this chapter, we will elaborate on our rationale for presenting yet another approach to well-posedness of partial differential equations. In particular, we will highlight conceptual differences between well established rationales and the one presented here. To start with, let us consider why we avoid the usual prominent role of the Laplacian in mathematical physics.

3.1 Where is the Laplacian?

Although, ab initio the equations of mathematical physics are mostly first order systems, it has been historically standard to reformulate them as second (or higher) order problems. A typical case is given by the system of Sect. 2.1. Let us start by assuming a simple block-diagonal structure for M_0, M_1:

$$\left(\partial_0 \begin{pmatrix} \alpha_0 & 0 \\ 0 & \alpha_1 \end{pmatrix} + \begin{pmatrix} \beta_0 & 0 \\ 0 & \beta_1 \end{pmatrix} + \begin{pmatrix} 0 & \operatorname{div} \\ \overset{\circ}{\operatorname{grad}} & 0 \end{pmatrix} \right) \begin{pmatrix} p \\ v \end{pmatrix} = \begin{pmatrix} f \\ 0 \end{pmatrix}. \tag{3.1.1}$$

Recall that the underlying spatial Hilbert space is $H = L^2(\Omega) \oplus L^2(\Omega)^3$ for some $\Omega \subseteq \mathbb{R}^3$. We may be thinking of this system as describing (approximately)—say—the propagation of acoustic waves (or heat, or porosity, or ...). In the acoustic interpretation p would be the pressure and v the velocity field of the wave motion. We have—for sake of definiteness—assumed here Dirichlet boundary conditions for p. Eliminating v from the system now yields

$$\partial_0 \alpha_0 p + \beta_0 p - \operatorname{div} (\partial_0 \alpha_1 + \beta_1)^{-1} \overset{\circ}{\operatorname{grad}} p = f. \tag{3.1.2}$$

© Springer Nature Switzerland AG 2020

R. Picard et al., *A Primer for a Secret Shortcut to PDEs of Mathematical Physics*,
Frontiers in Mathematics, https://doi.org/10.1007/978-3-030-47333-4_3

By differentiating with respect to time, we get a second-order-in-time problem of the form

$$\partial_0^2 \alpha_0 p + \beta_0 \partial_0 p - \text{div } \partial_0 \, (\partial_0 \alpha_1 + \beta_1)^{-1} \, \overset{\circ}{\text{grad}} \, p = \partial_0 f. \tag{3.1.3}$$

It is less than obvious how one would solve the resulting equation for the pressure distribution p, although it is obvious that well-posedness of the original first order problem requires merely (1.3.1) to hold.

Indeed, only for very particular materials can Eq. (3.1.3) actually be turned into a proper second order problem. It is also clear that the answer to the title of this section is: Well, there may not be any Laplacian at all!

Let us for simplicity assume that

$$\alpha_1 = r\mu$$

$$\beta_1 = s\mu$$

for some numbers $r \in [0, \infty[$, $s \in \mathbb{R}$, such that either $s \in]0, \infty[$ (if $r = 0$) or $s \in \mathbb{R}$ (if $r \neq 0$), and some continuous, selfadjoint, strictly positive definite $\mu: L^2(\Omega)^3 \to L^2(\Omega)^3$. Then (3.1.2) formally simplifies to a proper second order problem

$$\partial_0^2 r \alpha_0 p + (\beta_0 r + s\alpha_0) \, \partial_0 p + \beta_0 s p - \text{div } \mu^{-1} \overset{\circ}{\text{grad}} \, p = r \partial_0 f + s f. \tag{3.1.4}$$

In particular, we are finally at the point where we can say: Here is the Laplacian (for $\mu = 1$)!

In the case $r = 0$, we may assume without loss of generality that $s = 1$ and we have

$$\alpha_0 \partial_0 p + \beta_0 p - \text{div } \mu^{-1} \overset{\circ}{\text{grad}} \, p = g. \tag{3.1.5}$$

If $\alpha_0 = 0$ assumption (1.3.1) implies that $\beta_0 + \beta_0^*$ is strictly positive definite and we end up finally with the equation

$$\left(\beta_0 - \text{div } \mu^{-1} \overset{\circ}{\text{grad}} \right) p = f, \tag{3.1.6}$$

which in the usual partial differential equations classification would likely be called *elliptic*. We should, however, not forget that we are still in the time-dependent situation although there is no time derivative appearing. Nevertheless, accepting time as a parameter, (3.1.6) can be approached by standard elliptic methods. In fact, in this case the time dependence is of little importance: if we assume for example that $f(t, x) = \chi_{[0,\infty[}(t) g(x)$ then p would be of the same form. In contrast, our method, although designed to solve dynamic equations, also yields well-posedness for the corresponding first order system

$$\left(\begin{pmatrix} \beta_0 & 0 \\ 0 & \mu \end{pmatrix} + \begin{pmatrix} 0 & \mathrm{div} \\ \overset{\circ}{\mathrm{grad}} & 0 \end{pmatrix} \right) \begin{pmatrix} p \\ v \end{pmatrix} = \begin{pmatrix} f \\ 0 \end{pmatrix},$$

since the assumptions that both $\beta_0 + \beta_0^*$ and μ strictly positive definite have been imposed.

So, let us return to (3.1.5) and assume now that α_0 is continuous, selfadjoint and strictly positive definite. Then (3.1.5) would be called *parabolic* (a "heat equation"). Any parabolic method can be used to attack this problem or we could realize that (3.1.5) can be solved by considering

$$\left(\partial_0 \begin{pmatrix} \alpha_0 & 0 \\ 0 & 0 \end{pmatrix} + \begin{pmatrix} \beta_0 & 0 \\ 0 & \mu \end{pmatrix} + \begin{pmatrix} 0 & \mathrm{div} \\ \overset{\circ}{\mathrm{grad}} & 0 \end{pmatrix} \right) \begin{pmatrix} p \\ v \end{pmatrix} = \begin{pmatrix} f \\ 0 \end{pmatrix}$$

in the framework of our theory, that is, Theorem 1.3.2.

Finally, if $r > 0$ we may without loss of generality let $r = 1$ in (3.1.4) to get

$$\partial_0^2 \alpha_0 p + \alpha_* \partial_0 p + \alpha_{**} p - \mathrm{div} \, \mu^{-1} \overset{\circ}{\mathrm{grad}} \, p = g, \tag{3.1.7}$$

with

$$\alpha_* := \beta_0 + s\alpha_0, \alpha_{**} := \beta_0 s, g := \partial_0 f + sf,$$

which would be classified as a *hyperbolic* partial differential equation (a "wave equation"). If we ignore how the coefficients were defined, we can still derive a corresponding evolutionary system of our standard type. Indeed, applying ∂_0^{-1} to (3.1.7) we get

$$\partial_0 \alpha_0 p + \alpha_* p + \alpha_{**} \partial_0^{-1} p + \mathrm{div} \, v = \partial_0^{-1} g$$

with

$$v := -\partial_0^{-1} \mu^{-1} \overset{\circ}{\mathrm{grad}} \, p.$$

Written as a system, this is

$$\left(\partial_0 \begin{pmatrix} \alpha_0 & 0 \\ 0 & \mu \end{pmatrix} + \begin{pmatrix} \alpha_* & 0 \\ 0 & 0 \end{pmatrix} + \partial_0^{-1} \begin{pmatrix} \alpha_{**} & 0 \\ 0 & 0 \end{pmatrix} + \begin{pmatrix} 0 & \mathrm{div} \\ \overset{\circ}{\mathrm{grad}} & 0 \end{pmatrix} \right) \begin{pmatrix} p \\ v \end{pmatrix} = \begin{pmatrix} \partial_0^{-1} g \\ 0 \end{pmatrix}, \tag{3.1.8}$$

where the term marked in red can be considered as a small perturbation of the simple mate-

rial law operator $\begin{pmatrix} \alpha_0 & 0 \\ 0 & \mu \end{pmatrix} + \partial_0^{-1} \begin{pmatrix} \alpha_* & 0 \\ 0 & 0 \end{pmatrix}$, see also Theorem 1.3.8, with the corresponding

unperturbed system

$$\left(\partial_0 \begin{pmatrix} \alpha_0 & 0 \\ 0 & \mu \end{pmatrix} + \begin{pmatrix} \alpha_* & 0 \\ 0 & 0 \end{pmatrix} + \begin{pmatrix} 0 & \mathrm{div} \\ \overset{\circ}{\mathrm{grad}} & 0 \end{pmatrix} \right) \begin{pmatrix} p \\ v \end{pmatrix} = \begin{pmatrix} \partial_0^{-1} g \\ 0 \end{pmatrix}. \tag{3.1.9}$$

The routine, that is to say classical, way to approach (3.1.7) is also to turn it into an
evolution equation, but usually in quite a different way than (3.1.8). Guided by the standard
strategy for ordinary differential equations $\dot{p} := \partial_0 p$ is introduced as an additional
unknown leading to the system

$$\left(\partial_0 \begin{pmatrix} \alpha_0 & 0 \\ 0 & 1 \end{pmatrix} + \begin{pmatrix} \alpha_* & -\alpha_{**} \\ 0 & 0 \end{pmatrix} + \begin{pmatrix} 0 & \mathrm{div}\, \mu^{-1} \overset{\circ}{\mathrm{grad}} \\ 1 & 0 \end{pmatrix} \right) \begin{pmatrix} \dot{p} \\ -p \end{pmatrix} = \begin{pmatrix} g \\ 0 \end{pmatrix}. \tag{3.1.10}$$

The resulting formal spatial operator

$$\begin{pmatrix} 0 & \mathrm{div}\, \mu^{-1} \overset{\circ}{\mathrm{grad}} \\ 1 & 0 \end{pmatrix} \tag{3.1.11}$$

needs to be properly interpreted to be accessible by classical (e.g. semi-group) methods.
Indeed, if one analyzes this approach, it turns out that the entry 1 in the bottom left corner
of (3.1.11) is *not* the identity mapping and $\mathrm{div}\, \mu^{-1} \overset{\circ}{\mathrm{grad}}$ is *not* the $L^2(\Omega)$-selfadjoint
operator suggested by the notation. The subtleties are hidden in the choice of a quite
different Hilbert space setting. Let us begin with the 1, which is actually the inverse of
the continuous canonical embedding

$$\iota_{\mathrm{d}} : \mathrm{dom}\left(\mu^{-1/2} \overset{\circ}{\mathrm{grad}} \right) \to L^2(\Omega)$$

$$\varphi \mapsto \varphi. \tag{3.1.12}$$

Note that

$$\iota_{\mathrm{d}}^{-1} : \mathrm{dom}\left(\mu^{-1/2} \overset{\circ}{\mathrm{grad}} \right) \subseteq L^2(\Omega) \to \mathrm{dom}\left(\mu^{-1/2} \overset{\circ}{\mathrm{grad}} \right)$$

$$\varphi \mapsto \varphi$$

is *not* a continuous operator, but is closed and densely defined. We shall calculate its adjoint:[1]

Proposition 3.1.1 *With ι_d given in* (3.1.12), *we have*

$$\left(\iota_d^{-1}\right)^* : \mathrm{dom}\left(\left(\iota_d^{-1}\right)^*\right) \subseteq \mathrm{dom}\left(\mu^{-1/2}\mathrm{grad}\right) \to L^2\left(\Omega\right),$$

with

$$\left(\iota_d^{-1}\right)^* = \left(1 - \mathrm{div}\,\mu^{-1}\overset{\circ}{\mathrm{grad}}\right)\iota_d. \tag{3.1.13}$$

Proof For all $\varphi \in \mathrm{dom}\left(\mu^{-1/2}\overset{\circ}{\mathrm{grad}}\right)$, $\psi \in \mathrm{dom}\left(\left(\iota_d^{-1}\right)^*\right) \subseteq \mathrm{dom}\left(\mu^{-1/2}\overset{\circ}{\mathrm{grad}}\right)$,

$$\left\langle\varphi\left|\left(\iota_d^{-1}\right)^*\psi\right.\right\rangle_{L^2(\Omega)} = \left\langle\iota_d^{-1}\varphi\middle|\psi\right\rangle_{\mathrm{dom}\left(\mu^{-1/2}\overset{\circ}{\mathrm{grad}}\right)}$$

$$= \langle\varphi|\psi\rangle_{L^2(\Omega)} + \left\langle\mu^{-1/2}\overset{\circ}{\mathrm{grad}}\,\varphi\middle|\mu^{-1/2}\overset{\circ}{\mathrm{grad}}\,\psi\right\rangle_{L^2(\Omega)^3}.$$

Thus, we read off

$$\mu^{-1}\overset{\circ}{\mathrm{grad}}\,\psi \in \mathrm{dom}\,(\mathrm{div})$$

and

$$\mathrm{div}\,\mu^{-1}\overset{\circ}{\mathrm{grad}}\,\psi = \psi - \left(\iota_d^{-1}\right)^*\psi$$

for all $\psi \in \mathrm{dom}\left(\left(\iota_d^{-1}\right)^*\right)$, yielding the assertion. $\qquad\square$

So, interpreting (3.1.11) as

$$A + \begin{pmatrix} 0 & \iota_d \\ 0 & 0 \end{pmatrix}$$

---

[1]Recall that for a closed operator A, we always consider dom (A) as a Hilbert space with respect to its graph inner product. Thus, as Hilbert spaces dom (A) and dom (B) maybe different, whereas as sets they can be equal!

with

$$A := \begin{pmatrix} 0 & -\left(\iota_d^{-1}\right)^* \\ \iota_d^{-1} & 0 \end{pmatrix}$$

on $L^2\left(\Omega\right) \oplus \mathrm{dom}\left(\mu^{-1/2}\mathring{\mathrm{grad}}\right)$, we see that, despite apparently lacking the appropriate structure in (3.1.11) initially, (3.1.10) actually can be written in standard form[2]

$$\partial_0 M_0 + M_1 + A \tag{3.1.14}$$

with

$$M_0 = \begin{pmatrix} \alpha_0 & 0 \\ 0 & 1 \end{pmatrix}, \quad M_1 = \begin{pmatrix} \alpha_* & -\alpha_{**} + \iota_d \\ 0 & 0 \end{pmatrix}. \tag{3.1.15}$$

Note, however, that in contrast to the above formal "derivation" the requirements on the coefficient operators are quite different, since now we need

$$\alpha_0, \, \alpha_*: \, \mathrm{dom}\left(\mu^{-1/2}\mathring{\mathrm{grad}}\right) \to \mathrm{dom}\left(\mu^{-1/2}\mathring{\mathrm{grad}}\right), \tag{3.1.16}$$

and

$$\alpha_{**}: \, \mathrm{dom}\left(\mu^{-1/2}\mathring{\mathrm{grad}}\right) \to L^2\left(\Omega\right).$$

Note that (3.1.16) expresses higher regularity assumptions for the coefficients. Moreover, the positivity requirement is now to be valid in $\mathrm{dom}\left(\mu^{-1/2}\mathring{\mathrm{grad}}\right)$, that is, with respect to the graph inner product of $\mu^{-1/2}\mathring{\mathrm{grad}}$. The problems that may arise from this perspective are largely diminished by assuming $\alpha_0, \alpha_*, \alpha_{**}$ to be real numbers.

The curiosity goes further: from (3.1.13) we see that

$$\left|\iota_d^{-1}\right|^2 = 1 + \left|\mu^{-1/2}\mathring{\mathrm{grad}}\right|^2$$

---

[2]We have not encountered any other authors, who use this kind of set-up, being aware of the skew-selfadjointness of the underlying spatial operator A.

which implies

$$\left|\iota_d^{-1}\right| = \sqrt{1 + \left|\mu^{-1/2}\overset{\circ}{\mathrm{grad}}\right|^2},$$

$$= \left|\left|\mu^{-1/2}\overset{\circ}{\mathrm{grad}}\right| + i\right|.$$

By Theorem 2.2.3, we now have that A is unitarily congruent to

$$\begin{pmatrix} 0 & -\left|\iota_d^{-1}\right| \\ \left|\iota_d^{-1}\right| & 0 \end{pmatrix} = \begin{pmatrix} 0 & -\left|\left|\mu^{-1/2}\overset{\circ}{\mathrm{grad}}\right| + i\right| \\ \left|\left|\mu^{-1/2}\overset{\circ}{\mathrm{grad}}\right| + i\right| & 0 \end{pmatrix},$$

which in turn is by the same argument unitarily congruent to

$$\begin{pmatrix} 0 & i - \left|\mu^{-1/2}\overset{\circ}{\mathrm{grad}}\right| \\ i + \left|\mu^{-1/2}\overset{\circ}{\mathrm{grad}}\right| & 0 \end{pmatrix} = \begin{pmatrix} 0 & i \\ i & 0 \end{pmatrix} + \begin{pmatrix} 0 & -\left|\mu^{-1/2}\overset{\circ}{\mathrm{grad}}\right| \\ \left|\mu^{-1/2}\overset{\circ}{\mathrm{grad}}\right| & 0 \end{pmatrix}.$$

Taking $\begin{pmatrix} 0 & i \\ i & 0 \end{pmatrix}$ as part of the material law, we are left with

$$\begin{pmatrix} 0 & -\left|\mu^{-1/2}\overset{\circ}{\mathrm{grad}}\right| \\ \left|\mu^{-1/2}\overset{\circ}{\mathrm{grad}}\right| & 0 \end{pmatrix},$$

which is unitarily congruent to

$$\begin{pmatrix} 0 & \mathrm{div}\,\mu^{-1/2} \\ \mu^{-1/2}\overset{\circ}{\mathrm{grad}} & 0 \end{pmatrix} = \begin{pmatrix} 0 & 0 \\ 0 & \mu^{-1/2} \end{pmatrix}\begin{pmatrix} 0 & \mathrm{div} \\ \overset{\circ}{\mathrm{grad}} & 0 \end{pmatrix}\begin{pmatrix} 0 & 0 \\ 0 & \mu^{-1/2} \end{pmatrix}$$

on the domain of this operator in the Hilbert space $L^2(\Omega) \oplus \overline{\mu^{-1/2}\overset{\circ}{\mathrm{grad}}\left[L^2(\Omega)\right]}$.

Thus, we reconfirm that the system (3.1.14) and (3.1.15) is indeed essentially congruent to the acoustic case which, had we properly considered the original first order system, would have been clear to begin with.

3.2   Why Not Use Semi-Groups?

Since the standard way to solve systems such as (3.1.14) and (3.1.15), or, more abstractly, any system of the form

$$(\partial_0 M_0 + M_1 + A)\,u = f,$$

with M_0 continuous, selfadjoint and strictly positive definite, would be to use semi-groups (see e.g. [1,12,21] for some standard references), this may be a good time to compare these methods to what we propose here. We recall, compare Proposition 2.5.5, that by using $\sqrt{M_0^{-1}}$ as a congruence we may indeed assume without loss of generality that $M_0 = 1$ and consider

$$(\partial_0 + M_1 + A)\, u = f \tag{3.2.1}$$

instead. Thus, the issue in the semi-group approach is to see that $-(M_1 + A)$ generates a semi-group. In our case, where A is assumed to be skew-selfadjoint, we can actually simplify matters further by noting that, for $\varrho \in {]}0, \infty[$ sufficiently large, M_1 is a small perturbation and the semi-group generated by A can actually be given in terms of the spectral theorem for the selfadjoint operator $\frac{1}{i}A$ in the underlying Hilbert space H, which allows the definition of a family U of unitary operators in H parameterized over $\mathbb{R}$, where

$$U\,(t) = \exp\,(-t A) := \exp\left(-it\left(\frac{1}{i}A\right)\right),\ t \in \mathbb{R}.$$

With respect to composition, the set $U\,[\mathbb{R}] = \{U(t);\, t \in \mathbb{R}\}$ of unitary operators is an Abelian group with $U\,(0) = 1$ as unit element and

$$U\,(t)\,U\,(s) = U\,(t + s)\,,\ t, s \in \mathbb{R}.$$

Following again the ordinary differential equations trail, we could consider the idea of a fundamental solution, which we can derive here from the group $U\,[\mathbb{R}]$. Indeed, $\chi_{\mathbb{R}_{>0}}U$ given by

$$\left(\chi_{\mathbb{R}_{>0}}U\right)(t) := \chi_{\mathbb{R}_{>0}}\,(t)\,U\,(t)\,,\ t \in \mathbb{R},$$

clearly satisfies (at least on dom (A))

$$(\partial_0 + A)\,\chi_{\mathbb{R}_{>0}}U = 0 \text{ on } \mathbb{R} \setminus \{0\}\,. \tag{3.2.2}$$

Moreover, we have

$$\left(\chi_{\mathbb{R}_{>0}}U\right)(0+) = U\,(0) = 1, \tag{3.2.3}$$

which indeed makes $G := \chi_{\mathbb{R}_{>0}} U$ what in the theory of ordinary differential equations would be called a fundamental solution[3] from which the general solution of (3.2.1) can—under suitable assumptions on the data—be obtained by convolution. Since we have cut off half the real axis, $U|_{[0,\infty[}$ is merely a semi-group of unitary operators. To find a fundamental solution for $(\partial_0 + M_1 + A)$, we note that M_1 is a small perturbation of $\partial_0 + A$ as long as $\varrho \in \,]0, \infty[$ is sufficiently large. The result is again a fundamental solution G_*, that is,

$$\partial_0 \left(G_* \left(\cdot \right) u_0 - \chi_{\mathbb{R}_{>0}} u_0 \right) + M_1 G_*(\cdot) u_0 + A G_* \left(\cdot \right) u_0 = 0$$

for $u_0 \in \mathrm{dom}\,(A)$, inducing an operator family $W = G_*|_{[0,\infty[}$, which yields again a semi-group $W\,[[0, \infty[]$ of continuous linear operators. Understanding that the semi-group approach gives rise to fundamental solutions sheds light on the usefulness of this methods in the context of evolution equations.

The semi-group approach further leads to additional representation results for the solution in terms of convolutions with the fundamental solution. Moreover, due to the strongly continuous dependence of the semi-group on the parameter, that is the time variable, the semi-group representation leads to subtle regularity results. For example, for $u_0 \in \mathrm{dom}\,(A)$ our abstract evolutionary equation theory would merely yield

$$G_* \left(\cdot \right) u_0 - \chi_{\mathbb{R}_{>0}} u_0 \in H_{\varrho,0} \left(\mathbb{R}; H \right),$$

whereas the strong continuity of the semi-group would show in this case additionally that

$$G_* \left(\cdot \right) u_0 - \chi_{\mathbb{R}_{>0}} u_0 \in C \left(\mathbb{R}; H \right),$$

---

[3] Another way of expressing the Eqs. (3.2.2) and (3.2.3), would be to write

$$(\partial_0 + A)\, G = \delta$$

or if we wished to avoid introducing the Dirac-δ-distribution and to establish the idea of operators as solutions:

$$\partial_0 \left(G \left(\cdot \right) u_0 - \chi_{\mathbb{R}_{>0}} u_0 \right) + A G \left(\cdot \right) u_0 = 0$$

for $u_0 \in \mathrm{dom}\,(A)$. Note that

$$G \left(\cdot \right) u_0 - \chi_{\mathbb{R}_{>0}} u_0 = \left(\chi_{\mathbb{R}_{>0}} U \left(\cdot \right) \right) u_0 - \chi_{\mathbb{R}_{>0}} u_0$$

has no jump at the origin and indeed a (weak) time derivative in $H_{\varrho,0} \left(\mathbb{R}; H \right)$, which is $G \left(\cdot \right) A u_0 = A G \left(\cdot \right) u_0$.

where $C(\mathbb{R}; H)$ is the space of H-valued continuous functions on $\mathbb{R}$, and even

$$G_*|_{[0,\infty[}(\,\cdot\,)\,u_0 \in C_1\left([0,\infty[\,,H\right).$$

For example the associated "energy balance"[4] for pure initial value problems

---

[4]This may not be the "energy balance" that the reader may recall (for example for acoustic waves). Let us assume that $M_1 = 0$. Then (3.2.5) is just stating that

$$\frac{1}{2}\langle u|u\rangle_H = \frac{1}{2}|u|_H^2$$

is constant in time on $[0,\infty[$. In terms of the group generated by A this is to say that

$$t \mapsto \frac{1}{2}|u|_H^2\,(t) = \frac{1}{2}\left|\exp\left(-tA\right)u_0\right|_H^2$$

is a constant function, which is of course a feature of the group elements being unitary, and the constant is $\frac{1}{2}|u_0|_H^2$. Further, since $\exp\left(-tA\right)$ commutes with any other (Borel) function f of $\frac{1}{i}A$ in the sense of spectral theory,

$$[0,\infty[\,\ni\,t \mapsto \frac{1}{2}\left|f\left(\frac{1}{i}A\right)\exp\left(-tA\right)u_0\right|_H^2 = \frac{1}{2}\left|\exp\left(-tA\right)f\left(\frac{1}{i}A\right)u_0\right|_H^2 = \frac{1}{2}\left|f\left(\frac{1}{i}A\right)u_0\right|_H^2$$

is also a constant function for all $u_0 \in \mathrm{dom}\left(f\left(\frac{1}{i}A\right)\right)$. In particular, assuming that f is the identity on $\mathbb{R}$ and $u_0 \in \mathrm{dom}\,(A)$, we get

$$t \mapsto \frac{1}{2}\left|\frac{1}{i}Au\,(t)\right|_H^2 = \frac{1}{2}|Au\,(t)|_H^2 = \frac{1}{2}|Au_0|_H^2$$

on $[0,\infty[$. In our current example, (3.1.9) with $\alpha_0 = 1$, $\mu = 1$ and $\alpha_* = 0$, $A = \begin{pmatrix} 0 & \mathrm{div} \\ \overset{\circ}{\mathrm{grad}} & 0 \end{pmatrix}$ and so

$$\frac{1}{2}|\mathrm{div}\,v\,(t)|_{L^2(\Omega)}^2 + \frac{1}{2}\left|\overset{\circ}{\mathrm{grad}}\,p\,(t)\right|_{L^2(\Omega)^3}^2 = \frac{1}{2}\left|\begin{pmatrix}\mathrm{div}\,v\,(t)\\ \overset{\circ}{\mathrm{grad}}\,p\,(t)\end{pmatrix}\right|_H^2 = \frac{1}{2}\left|\begin{pmatrix}\mathrm{div}\,v_0\\ \overset{\circ}{\mathrm{grad}}\,p_0\end{pmatrix}\right|_H^2$$

for $t \in [0,\infty[$. From the first equation of (3.1.9) we know that $\partial_0 p = -\,\mathrm{div}\,v$ on $]0,\infty[$. Thus,

$$\frac{1}{2}|\partial_0 p\,(t)|_{L^2(\Omega)}^2 + \frac{1}{2}\left|\overset{\circ}{\mathrm{grad}}\,p\,(t)\right|_{L^2(\Omega)^3}^2 = \frac{1}{2}\left|\begin{pmatrix}\mathrm{div}\,v_0\\ \overset{\circ}{\mathrm{grad}}\,p_0\end{pmatrix}\right|_H^2 \tag{3.2.4}$$

for $t \in\,]0,\infty[$. By the continuity properties of the group generated by A, (3.2.4) also holds for $t = 0$ so that

$$\frac{1}{2}\left(|\partial_0 p\,(t)|_{L^2(\Omega)}^2 + \left|\overset{\circ}{\mathrm{grad}}\,p\,(t)\right|_{L^2(\Omega)^3}^2\right) = \frac{1}{2}\left(|\partial_0 p_0|_{L^2(\Omega)}^2 + \left|\overset{\circ}{\mathrm{grad}}\,p_0\right|_{L^2(\Omega)^3}^2\right)$$

$$\frac{1}{2} \langle u|u \rangle_H (T) + \int_\tau^T \langle u|M_1 u \rangle_H = \frac{1}{2} \langle u|u \rangle_H (\tau) \qquad (3.2.5)$$

can now be understood as an equality between continuous functions on any non-empty subinterval $[\tau, T] \subseteq [0, \infty[$. The gain in regularity information stemming from the availability of a fundamental solution facilitates, similar to the role of fundamental solutions in partial differential equations, the transfer to a general Banach space setting. We further note that in [81] the link from evolutionary equations to C_0-semigroups has been developed. In fact, it turns out that it is possible to associate a strongly continuous semi-group with an evolutionary equation only if suitable regularity requirements are met.

If we consider problems in the class we have studied, it should be noted that, even if A (or $M_1 + A$) has a semi-group associated with it, this semi-group is less than useful if M_0 is *not* strictly positive definite.

Indeed, although, we have not highlighted the issue, we can consider material laws which make the equations change their classical "type", see e.g. [13, 90] for 1+1-dimensional problems and to [70, Remark 6.2] for an example in control theory and to [63, p. 765] for a non-autonomous situation. This is clearly a situation, where a fundamental solution could be derived only in very exceptional cases see [3]. For our theory, such problems can be dealt with no extra effort. The only requirement is assumption (1.3.1). So, even in our simple block-diagonal case

$$\left(\partial_0 \begin{pmatrix} \alpha_0 & 0 \\ 0 & \alpha_1 \end{pmatrix} + \begin{pmatrix} \beta_0 & 0 \\ 0 & \beta_1 \end{pmatrix} + \begin{pmatrix} 0 & \mathrm{div} \\ \overset{\circ}{\mathrm{grad}} & 0 \end{pmatrix} \right) \begin{pmatrix} p \\ v \end{pmatrix} = \begin{pmatrix} f \\ 0 \end{pmatrix} \qquad (3.2.6)$$

we could allow α_0 or α_1 to vanish in some regions. Suppose there are three disjoint measurable subdomains $\Omega_1, \Omega_2, \Omega_3$ of Ω with $\Omega_1 \cup \Omega_2 \cup \Omega_3$ dense in Ω, such that, for simplicity, the coefficients commute with the characteristic functions (as multiplication operators) $\chi_{\Omega_1}, \chi_{\Omega_2}, \chi_{\Omega_3}$. Then, let α_0 be strictly positive definite on $L^2 (\Omega_1)$ but $\alpha_1 = 0$ on $L^2 (\Omega_1)^3$. Let $\alpha_0 = 0$ on $L^2 (\Omega_2)$ and $\alpha_1 = 0$ on $L^2 (\Omega_2)^3$. Finally, let α_0 and α_1 be strictly positive definite on $L^2 (\Omega_3)$ and $L^2 (\Omega_3)^3$, respectively. Then, (3.2.6) has wave behavior in Ω_3, dissipative behavior in Ω_1 and quasi-static behavior in Ω_2.

The problem is compounded for more general coefficient operators M_0, M_1 and possible additional small perturbation terms, compare (2.5.9) and (2.5.10), such as so-called metamaterials, which have attracted a strong interest from the engineering community in recent years.

---

for $t \in [0, \infty[$, which is the more common form of energy conservation in the context of acoustic waves as the sum of "kinetic energy" $\frac{1}{2} |\partial_0 p (t)|^2_{L^2 (\Omega)}$ and "potential energy" $\frac{1}{2} \left| \overset{\circ}{\mathrm{grad}} p (t) \right|^2_{L^2 (\Omega)^3}$.

3.3   What About Other Types of Equations?

Amongst the many increasingly complex mathematical models (i.e. systems of equations) being introduced to describe real world phenomena, there are many which are not obviously of the evolutionary form discussed here. We have looked at some of the more accessible coupled system examples in the previous section.

On the other hand, the assumption (1.3.1) appears to be close to optimal in the sense that, if we consider material law operators for which this assumption fails, for instance M_0 is not selfadjoint or $\varrho M_0 + (1/2)\left(M_1 + M_1^*\right)$ is merely non-negative for all sufficiently large $\varrho \in \,]0, \infty[$ and has a non-trivial null space, we can easily find (simple $1 + 1$-dimensional) counter-examples, where either well-posedness or causality fails to hold (even for the case $A = 0$), these being essential properties for a mathematical model of a natural phenomenon.

The strong restriction (1.3.1) is due to the fact that we are, for non-zero M_0, dealing with a sum[5] of two discontinuous operators $\partial_0 M_0$ and $M_1 + A$:

$$\partial_0 M_0 + M_1 + A = (\partial_0 M_0) + (M_1 + A),$$

which under more general circumstances may be rather trivial: for instance the domain could consist only of 0.

More often than not, however, equations have been developed in a form which is not conducive to our approach. The reason is that, due to their motivation and derivation, the equations often end up in a form

$$W\left(\partial_0 M_0 + M_1 + A\right) V \tag{3.3.1}$$

with W, V bijections. Unfortunately, the factorization (3.3.1) is often by no means obvious, but usually needs to be found, see also the introductory part of Sect. 2.6.2. Moreover, the calculations are commonly quite formal, so that the term "bijection" is used without much examination of domain issues. A typical example can be taken from (3.1.9), which we may formally write as

$$\begin{pmatrix} \partial_0 \alpha_0 + \alpha_* \operatorname{div} \\ \overset{\circ}{\operatorname{grad}} \quad \partial_0 \mu \end{pmatrix} \begin{pmatrix} p \\ v \end{pmatrix} = \begin{pmatrix} \partial_0^{-1} g \\ 0 \end{pmatrix}.$$

---

[5]There is a well-developed theory for sums of unbounded operators, see e.g. [7, 10]. The criteria are, however, not easily manageable and go far beyond the complexity of our simple setup. In hindsight, it seems indeed rather wasteful and misleading to further the impression that considerations of such sophistication are needed to understand the standard evolutionary problems of engineering and mathematical physics.

The dramatic simplification in our setup is due to the strict positive definiteness of ∂_0, to the skew-selfadjointness of A and the fact that ∂_0 and A commute.

By (formal) row-operations we get

$$\begin{pmatrix} \partial_0^2 \alpha_0 + \alpha_* \partial_0 - \operatorname{div} \mu^{-1} \overset{\circ}{\operatorname{grad}} & 0 \\ \overset{\circ}{\operatorname{grad}} & \partial_0 \mu \end{pmatrix} \begin{pmatrix} p \\ v \end{pmatrix} = \begin{pmatrix} g \\ 0 \end{pmatrix},$$

and so

$$W = \begin{pmatrix} \partial_0 & -\operatorname{div} \mu^{-1} \\ 0 & 1 \end{pmatrix}, \quad V = 1.$$

Although, this has generated the second order wave equation (and can actually be made rigorous, as W can indeed be established as a bijection between appropriate spaces), the original structural information, that M_0 is selfadjoint and $\partial_0 M_0 + M_1 + A$ satisfies, due to (1.3.1), the numerical range condition

$$w(\partial_0 M_0 + M_1 + A) \subset [c_0, \infty[$$

for some $c_0 \in \,]0, \infty[$, is totally obscured ($w(T)$ denotes the numerical range of an operator T).[6]

---

[6]If we had considered second order in time equations, then the numerical range condition is more involved. Indeed, note that

$$\partial_0^2 = ((\partial_0 - \varrho) + \varrho)^2$$
$$= \varrho^2 + (\partial_0 - \varrho)^2 + 2\varrho (\partial_0 - \varrho)$$

and so, since $\partial_0 - \varrho$ is skew-selfadjoint,

$$\mathfrak{Re}\, \partial_0^2 = \varrho^2 + (\partial_0 - \varrho)^2$$
$$= \varrho^2 - \left(\frac{1}{i}(\partial_0 - \varrho)\right)^2,$$
$$\mathfrak{Im}\, \partial_0^2 = 2\varrho \frac{1}{i}(\partial_0 - \varrho).$$

Since the spectrum $\sigma\left(\frac{1}{i}(\partial_0 - \varrho)\right)$ is all of $\mathbb{R}$ we get

$$w\left(\partial_0^2\right) = \sigma\left(\partial_0^2\right) = \left\{\varrho^2 - r^2 + i2\varrho r \mid r \in \mathbb{R}\right\}$$
$$= \left\{\varrho^2 - \frac{1}{4}\varrho^{-2}s^2 + is \mid s \in \mathbb{R}\right\},$$

which is a parabola opening to the left, symmetric around the real axis, based at ϱ^2.

As well as such "bad transformations" which fall, at least formally, into the category of equivalences, there are also "good transformations": congruences, that is, transformations with $V = W^*$. Congruences W preserve (1.3.1) (with a possibly different positive definiteness constant) and the equation

$$(\partial_0 M_0 + M_1 + A)\, u = f$$

translates to

$$\left(\partial_0 W M_0 W^* + W M_1 W^* + W A W^*\right) v = W f,$$

where

$$v = \left(W^*\right)^{-1} u,$$

from which u can be recovered as $u = W^* v$. A particularly convenient case is of course the case of unitary congruence, where W is unitary and so $v = W u$. We have encountered this at several occasions. The next example will give another, more concrete, illustration of this case.

Example 3.3.1   Consider the transport equation

$$\left(\partial_0 + \overset{\#}{\partial_1}\right) u = f \tag{3.3.2}$$

where $\overset{\#}{\partial_1}$ denotes differentiation ∂_1 in $L^2\left(\left]-1, 1\right[\right)$ subject to the constraint

$$\mathrm{dom}\left(\overset{\#}{\partial_1}\right) = \{u \in \mathrm{dom}\left(\partial_1\right) \mid u\left(-1\right) = u\left(1\right)\},$$

that is, with periodic boundary condition. It is well-known and easy to check that $\overset{\#}{\partial_1}$ is skew-selfadjoint and so (3.3.2) falls into our problem class. Indeed, we observe that $L^2\left(\left]-1, 1\right[\right)$ can be orthogonally decomposed into even and odd parts

$$L^2\left(\left]-1, 1\right[\right) = L^2_{\mathrm{even}}\left(\left]-1, 1\right[\right) \oplus L^2_{\mathrm{odd}}\left(\left]-1, 1\right[\right).$$

If ι_{even} and ι_{odd} denote the canonical isometric embeddings of $L^2_{\mathrm{even}}\left(\left]-1, 1\right[\right)$ and $L^2_{\mathrm{odd}}\left(\left]-1, 1\right[\right)$ into $L^2\left(\left]-1, 1\right[\right)$, respectively, then we have the unitary mapping

$$\begin{pmatrix} \iota^*_{\mathrm{odd}} \\ \iota^*_{\mathrm{even}} \end{pmatrix} : L^2\left(\left]-1, 1\right[\right) \to L^2_{\mathrm{odd}}\left(\left]-1, 1\right[\right) \oplus L^2_{\mathrm{even}}\left(\left]-1, 1\right[\right),$$

with

$$\left(\iota_{\text{odd}}^* f \right)(x) = \frac{1}{2} \left(f(x) - f(-x) \right),$$

$$\left(\iota_{\text{even}}^* f \right)(x) = \frac{1}{2} \left(f(x) + f(-x) \right)$$

for all $f \in L^2\left( ]-1, 1[\right)$ and almost all $x \in \,]-1, 1[$. Noting that derivatives of odd functions are even and vice versa, we get

$$\left(\partial_0 + \begin{pmatrix} \iota_{\text{odd}}^* \\ \iota_{\text{even}}^* \end{pmatrix} \overset{\#}{\partial_1} \begin{pmatrix} \iota_{\text{odd}} & \iota_{\text{even}} \end{pmatrix} \right) \begin{pmatrix} \iota_{\text{odd}}^* u \\ \iota_{\text{even}}^* u \end{pmatrix} = \begin{pmatrix} \iota_{\text{odd}}^* f \\ \iota_{\text{even}}^* f \end{pmatrix},$$

$$\begin{pmatrix} 0 & \iota_{\text{odd}}^* \overset{\#}{\partial_1} \iota_{\text{even}} \\ \iota_{\text{even}}^* \overset{\#}{\partial_1} \iota_{\text{odd}} & 0 \end{pmatrix}.$$

Here $\iota_{\text{even}}, \iota_{\text{odd}}, \iota_{\text{even}}^*, \iota_{\text{odd}}^*$ are acting as identities and

$$\text{dom} \left(\overset{\#}{\partial_1} \right) = \left\{ u \in \text{dom}\,(\partial_1) \mid \left(\iota_{\text{odd}}^* u \right)(\pm 1) = 0 \right\}.$$

Hence, we may write

$$\begin{pmatrix} \iota_{\text{even}}^* \\ \iota_{\text{odd}}^* \end{pmatrix} \overset{\#}{\partial_1} \begin{pmatrix} \iota_{\text{odd}} & \iota_{\text{even}} \end{pmatrix} = \begin{pmatrix} 0 & \partial_1 \\ \overset{\circ}{\partial} & 0 \end{pmatrix},$$

if it is understood that the underlying space is

$$L_{\text{odd}}^2 \left( ]-1, 1[\right) \oplus L_{\text{even}}^2 \left( ]-1, 1[\right).$$

This shows that the periodic boundary condition is indeed just a Dirichlet boundary condition in disguise. Transforming back we obtain the following characterization[7]:

$$\overset{\#}{\partial_1} = \begin{pmatrix} \iota_{\text{odd}} & \iota_{\text{even}} \end{pmatrix} \begin{pmatrix} 0 & \partial_1 \\ \overset{\circ}{\partial} & 0 \end{pmatrix} \begin{pmatrix} \iota_{\text{odd}}^* \\ \iota_{\text{even}}^* \end{pmatrix} \tag{3.3.3}$$

$$= \iota_{\text{odd}} \partial_1 \iota_{\text{even}}^* + \iota_{\text{even}} \overset{\circ}{\partial_1} \iota_{\text{odd}}^*.$$

---

[7]This construction can be lifted to the higher dimensional situation provided Ω has the appropriate symmetry properties.

This insight has been successfully applied to one-dimensional so-called port-Hamiltonian systems, see [80].

3.4   What About Other Boundary Conditions?

So far we have considered only concrete examples involving boundary conditions generated by differential operators D and D_* which are formally adjoint to each other on (component-wise) smooth functions, which vanish outside of a compact subset of Ω. Thus taking any closed extension of D will induce an adjoint C^* which extends the differential operator D_*, so that

$$A = \begin{pmatrix} 0 & -C^* \\ C & 0 \end{pmatrix}$$

is a skew-selfadjoint extension of the skew-symmetric $\begin{pmatrix} 0 & -D_* \\ D & 0 \end{pmatrix}$. This situation describes what it means to have 'separated' boundary conditions. It may be somewhat surprising that so-called coupled boundary condition may also fall into our standard class. They can be encoded by allowing C to contain boundary operators. To avoid excessive complication, we restrict our discussion to one space-dimension; for the general ideas see [69, 70, 78, 80], which discuss higher-dimensional cases and the issue of an irregular boundary.

Example 3.4.1 Consider formally the 1-dimensional case of (3.1.9) with no boundary condition

$$\left(\partial_0 \begin{pmatrix} \alpha_0 & 0 \\ 0 & \mu \end{pmatrix} + \begin{pmatrix} \alpha_* & 0 \\ 0 & 0 \end{pmatrix} + \begin{pmatrix} 0 & \partial_1 \\ \partial_1 & 0 \end{pmatrix} \right) \begin{pmatrix} p \\ v \end{pmatrix} = \begin{pmatrix} f \\ 0 \end{pmatrix}, \tag{3.4.1}$$

where the underlying spatial Hilbert space is $L^2(]-1, 1[) \oplus L^2(]-1, 1[)$. To implement boundary conditions we consider a problem, which is again of standard form,[8]

---

[8]Introducing a skew-selfadjoint spatial operator is not just an interesting "trick". In the spirit of footnote 2, this way we get an energy balance law for any solution $\left(\begin{pmatrix} p \\ v \\ \beta \end{pmatrix} \right)$.

$$\left(\partial_0 \left(\begin{array}{cc} \alpha_0 & \begin{pmatrix} 0 & 0 \end{pmatrix} \\ \begin{pmatrix} 0 \\ 0 \end{pmatrix} & \begin{pmatrix} \mu & 0 \\ 0 & \sigma \end{pmatrix} \end{array} \right) + \left(\begin{array}{cc} \alpha_* & \begin{pmatrix} 0 & 0 \end{pmatrix} \\ \begin{pmatrix} 0 \\ 0 \end{pmatrix} & \begin{pmatrix} 0 & 0 \\ 0 & 0 \end{pmatrix} \end{array} \right) \right.$$

$$\left. + \left(\begin{array}{cc} 0 & -\begin{pmatrix} \partial_1 \\ \gamma \end{pmatrix}^* \\ \begin{pmatrix} \partial_1 \\ \gamma \end{pmatrix} & \begin{pmatrix} 0 & 0 \\ 0 & 0 \end{pmatrix} \end{array} \right) \right) \begin{pmatrix} p \\ v \\ \beta \end{pmatrix} = \begin{pmatrix} f \\ 0 \\ h \end{pmatrix} \qquad (3.4.2)$$

where we have kept the material law simple. The additional component γ is

$$\gamma : \operatorname{dom}(\partial_1) \subseteq L^2\,(]-1, 1[) \to \mathbb{R}^2,$$

$$\varphi \mapsto \begin{pmatrix} \varphi\,(-1) \\ \varphi\,(1) \end{pmatrix},$$

where the values at -1 and 1 should be understood as limits from the right and from the left, respectively. The term γf represents the boundary trace of f. The additional right-hand side $h = \begin{pmatrix} h_1 \\ h_2 \end{pmatrix}$ is in $H_{\varrho,0}\,(\mathbb{R}, \mathbb{R}^2)$, that is, $h_k \in H_{\varrho,0}\,(\mathbb{R})$, $k \in \{1, 2\}$. Assuming as always (1.3.1) with

$$M_0 = \left(\begin{array}{cc} \alpha_0 & \begin{pmatrix} 0 & 0 \end{pmatrix} \\ \begin{pmatrix} 0 \\ 0 \end{pmatrix} & \begin{pmatrix} \mu & 0 \\ 0 & \sigma \end{pmatrix} \end{array} \right), \quad M_1 = \left(\begin{array}{cc} \alpha_* & \begin{pmatrix} 0 & 0 \end{pmatrix} \\ \begin{pmatrix} 0 \\ 0 \end{pmatrix} & \begin{pmatrix} 0 & 0 \\ 0 & 0 \end{pmatrix} \end{array} \right),$$

we get well-posedness and causality.

Now, noting that

$$\begin{pmatrix} \mathring{\partial}_1 \\ 0 \end{pmatrix} \subseteq \begin{pmatrix} \partial_1 \\ \gamma \end{pmatrix},$$

we obtain

$$\begin{pmatrix} \partial_1 \\ \gamma \end{pmatrix}^* \subseteq \begin{pmatrix} -\partial_1 & 0 \end{pmatrix}.$$

Thus, we have from (3.4.2)

$$
\left(\partial_0 \begin{pmatrix} \alpha_0 & \begin{pmatrix} 0 & 0 \\ \end{pmatrix} \\ \begin{pmatrix} 0 \\ 0 \end{pmatrix} & \begin{pmatrix} \mu & 0 \\ 0 & \sigma \end{pmatrix} \end{pmatrix} + \begin{pmatrix} \alpha_* & \begin{pmatrix} 0 & 0 \\ \end{pmatrix} \\ \begin{pmatrix} 0 \\ 0 \end{pmatrix} & \begin{pmatrix} 0 & 0 \\ 0 & 0 \end{pmatrix} \end{pmatrix} \right.
$$
$$
\left. + \begin{pmatrix} 0 & \begin{pmatrix} \partial_1 & 0 \\ \end{pmatrix} \\ \begin{pmatrix} \partial_1 \\ \gamma \end{pmatrix} & \begin{pmatrix} 0 & 0 \\ 0 & 0 \end{pmatrix} \end{pmatrix} \right) \begin{pmatrix} p \\ v \\ \beta \end{pmatrix} = \begin{pmatrix} f \\ 0 \\ h \end{pmatrix}. \qquad (3.4.3)
$$

Focusing on the first two rows, we see that we have recovered (3.4.1). Moreover, since (see Proposition 2.4.3)

$$
\begin{pmatrix} \partial_1 \\ \gamma \end{pmatrix}^* \subseteq \begin{pmatrix} \partial_1^\diamond & \gamma^\diamond \end{pmatrix},
$$

we have

$$
\begin{pmatrix} \partial_1 \\ \gamma \end{pmatrix}^* \begin{pmatrix} v \\ \beta \end{pmatrix} = \partial_1^\diamond v + \gamma^\diamond \beta = -\partial_1 v.
$$

Analyzing the last equality, we see that it implies

$$
\begin{aligned}
0 &= \langle \varphi | \partial_1 v \rangle_{L^2(]-1,1[)} + \langle \varphi | \partial_1^\diamond v + \gamma^\diamond \beta \rangle_{L^2(]-1,1[)} \\
&= \langle \varphi | \partial_1 v \rangle_{L^2(]-1,1[)} + \langle \partial_1 \varphi | v \rangle_{L^2(]-1,1[)} + \langle \gamma \varphi | \beta \rangle_{\mathbb{R}^2} \\
&= \varphi (1-0) v (1-0) - \varphi (-1+0) v (-1+0) + \langle \gamma \varphi | \beta \rangle_{\mathbb{R}^2} \\
&= \langle \gamma \varphi | \gamma v \rangle_{\mathbb{R}^2} + \langle \gamma \varphi | \beta \rangle_{\mathbb{R}^2}
\end{aligned}
$$

for all $\varphi \in \mathrm{dom}\,(\partial_1)$. Since γ is obviously onto, this yields

$$
\beta = -\gamma v, \qquad (3.4.4)
$$

which finally explains the role of this newly introduced unknown β. We can now inspect the last row of (3.4.3), which is

$$
\partial_0 \sigma \beta + \gamma p = h.
$$

Substituting (3.4.4) yields

$$
-\partial_0 \sigma \gamma v + \gamma p = h
$$

or, assuming that $\sigma : \mathbb{R}^2 \to \mathbb{R}^2$ is also block-diagonal, that is, $\sigma = \begin{pmatrix} \sigma_1 & 0 \\ 0 & \sigma_2 \end{pmatrix}$, $\sigma_k \in]0, \infty[$, $k \in \{1, 2\}$,

$$-\partial_0 \sigma_1 v\left(\,\cdot\,, 1 - 0\right) + p\left(1 - 0\right) = h_1,$$
$$-\partial_0 \sigma_2 v\left(\,\cdot\,, -1 + 0\right) + p\left(-1 + 0\right) = h_2,$$

which are ordinary differential equations on the boundary.

3.5   Why All This Functional Analysis?

It is a well-known fact, that many people working with partial differential equations have a deep hatred of functional analysis, usually denounced as "soft" analysis as opposed to "hard" analysis. Of course, this is sometimes a tongue-in-cheek attitude, but still somewhat irritating, since functional analysis plays such an important role in the field, and also there can be no objection to working "smart" instead of "hard".

Indeed, a functional analytical setting usually leads to more and deeper insight. The abstraction may also simplify arguments and reduce complexity. To give an example, it seems hard to believe that the Picard–Lindelöf theorem or various iteration schemes had been fully understood before the advent of Banach's fixed point theorem, which served to unify dramatically these rather diverse topics. Of course, people seeking their challenge in coping with complexity may be disappointed by such a radical simplification, as much as people trying to break open a door to get into the other room, will be annoyed[9] if someone finds out that the door is unlocked and just opens the other way.

Our approach reduces a number of issues in partial differential equations to a rather elementary, easy to check set of assumptions, requiring nothing but a solid foundation in functional analysis of Hilbert spaces to be fully understood. In the light of the quite general applicability of our simple setting, it seems to be a sound investment to acquire or to recover this foundation.

Moreover, we think that the easily checked assumptions in our solution theory should be attractive to researchers in applied areas such as mathematical physics or engineering rather than having to address more sophisticated sets of assumptions required for other (abstract) solution strategies.

---

[9]Some will find it a good solution of the problem to look for another door, which is really locked, to deploy their door-breaking skills there instead!

It is also noteworthy that the focus on regularity in the field of partial differential equations might mislead researchers from applied fields. Though our approach disregards regularity theory almost entirely, it is particularly useful for solving real life problems numerically (see [13, 14] for numerical treatments of evolutionary equations), where any kind of regularity gain cannot be expected from the equation at hand.

Two Supplements for the Toolbox

A

A.1   Mothers and Their Descendants

We start out with notions developed and studied in [60].

Definition A.1.1 Let $A : \mathrm{dom}(A) \subseteq H \to H$ be a skew-selfadjoint operator in a Hilbert space H. Then any skew-selfadjoint operator of the form

$$\overline{V^*AV},$$

where $V : X \to H$ is continuous and linear, X a Hilbert space, is called a *descendant of* A. In relation to $\overline{V^*AV}$ the operator A is referred to as its *mother*.

If V is a bijection, then V^*AV is always skew-selfadjoint with its natural domain $V^{-1}[\mathrm{dom}(A)]$. More interesting and complicated is the name-motivating case, where V is not invertible.

Definition A.1.2 Let $C : \mathrm{dom}(C) \subseteq H_0 \to H_1$ be linear, closed and densely defined and $B : H_0 \to X$ a continuous linear mapping, X, H_0, H_1 Hilbert spaces. We say B is *compatible with* C if CB^* is densely defined (in X).

Theorem A.1.3 *Let $C : \mathrm{dom}(C) \subseteq H_0 \to H_1$ be linear, closed and densely defined and $B : H_0 \to X$ a continuous linear mapping, X, H_0, H_1 Hilbert spaces. Moreover, let B be compatible with C. Then*

$$\left(CB^*\right)^* = \overline{BC^*}.$$

© Springer Nature Switzerland AG 2020
R. Picard et al., *A Primer for a Secret Shortcut to PDEs of Mathematical Physics*,
Frontiers in Mathematics, https://doi.org/10.1007/978-3-030-47333-4

Proof Note that C^* is densely defined since C is closed. Thus, $(BC^*)^*$ is an operator and

$$CB^* \subseteq (BC^*)^*.$$

Let $u \in \mathrm{dom}\big((BC^*)^*\big)$. Then for $v \in \mathrm{dom}(BC^*) = \mathrm{dom}(C^*)$:

$$\big\langle v \big| (BC^*)^* u \big\rangle_{H_1} = \langle BC^* v | u \rangle_X$$
$$= \langle C^* v | B^* u \rangle_{H_0}.$$

We read off that

$$B^* u \in \mathrm{dom}\big(C^{**}\big) = \mathrm{dom}(C)$$

and

$$CB^* u = (BC^*)^* u.$$

Consequently, we have

$$CB^* = (BC^*)^*$$

and so, since CB^* is densely defined,

$$(CB^*)^* = (BC^*)^{**} = \overline{BC^*}. \qquad \Box$$

This observation leads to the following result.

Theorem A.1.4 *Let* $C : \mathrm{dom}(C) \subseteq H_0 \to H_1$ *be closed and densely defined,* H_0, H_1 *Hilbert spaces so that*

$$A = \begin{pmatrix} 0 & -C^* \\ C & 0 \end{pmatrix}$$

is skew-selfadjoint in $H_0 \oplus H_1$. *Let*

$$V := \begin{pmatrix} U & 0 \\ 0 & W \end{pmatrix}$$

where $U : X \to H_0$, $W : Y \to H_1$ are continuous linear mappings, X, Y Hilbert spaces. Then, if W^* is compatible with C^* and U is a bijection, $\overline{V^*AV}$ is a descendant of A. In particular,

$$\left(\overline{V^*AV}\right)^* = \left(V^*AV\right)^* = -\overline{V^*AV} = -V^*AV. \qquad (A.1.1)$$

Proof Observing that

$$V = \begin{pmatrix} U & 0 \\ 0 & 1 \end{pmatrix} \begin{pmatrix} 1 & 0 \\ 0 & W \end{pmatrix},$$

and U is a bijection, we see that we may assume without loss of generality[1] that $U = 1$. Consequently, we have that

$$V^*AV = \begin{pmatrix} 0 & -C^*W \\ W^*C & 0 \end{pmatrix}$$

and so

$$\overline{V^*AV} = \overline{\begin{pmatrix} 0 & -C^*W \\ W^*C & 0 \end{pmatrix}} = \begin{pmatrix} 0 & -C^*W \\ W^*C & 0 \end{pmatrix} = V^*AV.$$

Moreover, since W^* is compatible with C^*, Theorem A.1.3 yields

$$\left(\overline{V^*AV}\right)^* = (V^*AV)^* = \begin{pmatrix} 0 & (W^*C)^* \\ -(C^*W)^* & 0 \end{pmatrix} = \begin{pmatrix} 0 & C^*W \\ -W^*C & 0 \end{pmatrix} = -V^*AV,$$

which proves the skew-selfadjointness of $\overline{V^*AV}$ and Eq. (A.1.1). □

Remark A.1.5

(1) The process of constructing descendants can be repeated, but may depend on the order in which the steps are carried out.

---

[1] Note that

$$(CU)^* = U^*C^*$$

and so this factorization simply means that the role of C is played by CU instead.

(2) A natural Hilbert space in which to consider $\overline{V^*AV}$ is the range space of V^*, that is,

$$\overline{V^*[H_0 \oplus H_1]} = X \oplus \overline{W^*[H_1]};$$

instead we consider the operator

$$\iota(V^*)^* \overline{V^*AV} \iota(V^*)$$

(recall the notation from Sect. 2.2.1). In this perspective, if we assume that W^* has closed range, in which case V^* has closed range, the positive definiteness condition of a mother operator of the form $\partial_0 M_0 + M_1 + A$ carries over to the descendant

$$\partial_0 \tilde{M}_0 + \tilde{M}_1 + \tilde{A}$$

with

$$\tilde{A} := \iota(V^*)^* \overline{V^*AV} \iota(V^*)$$
$$\tilde{M}_k := \iota(V^*)^* V^*M_k V \iota(V^*), \ k \in \{0, 1\}.$$

A.2 Abstract grad-div-Systems

We rephrase an observation from [69]. The standard form of the skew-selfadjoint operator A as

$$\begin{pmatrix} 0 & -C^* \\ C & 0 \end{pmatrix}$$

is, in the case of Sect. 2.1, generated by the gradient:

$$C = \text{grad} = \begin{pmatrix} \partial_1 \\ \partial_2 \\ \partial_3 \end{pmatrix} : \text{dom}(\text{grad}) \subseteq L^2(\Omega) \to L^2(\Omega)^3,$$

$$\varphi \mapsto \begin{pmatrix} \partial_1 \varphi \\ \partial_2 \varphi \\ \partial_3 \varphi \end{pmatrix},$$

where

$$\partial_k : \text{dom}(\text{grad}) \subseteq L^2(\Omega) \to L^2(\Omega), \ k \in \{1, 2, 3\},$$

are densely defined linear operators (clearly not closed, although grad is closed). The idea of abstract grad-div-systems is to replace the role of partial derivatives by general operators (and extend to higher dimensions).

Definition Let $H_0, \ldots, H_n$ be Hilbert spaces and

$$
C = \begin{pmatrix} C_1 \\ \vdots \\ C_n \end{pmatrix} : \operatorname{dom}(C) \subseteq H_0 \to H_1 \oplus \cdots \oplus H_n,
$$

$$
\varphi \mapsto \begin{pmatrix} C_1 \varphi \\ \vdots \\ C_n \varphi \end{pmatrix},
$$

be a closed, densely defined, linear operator. Then we call the system

$$
\partial_0 M_0 + M_1 + A,
$$

for linear continuous operators $M_0, M_1 : H \to H$, $H := H_0 \oplus \ldots \oplus H_n$ and

$$
A = \begin{pmatrix} 0 & -C^* \\ C & 0 \end{pmatrix}
$$

an *abstract* grad-div-*system generated by* C.

Remark A.2.1 For C as above, the operators

$$
C_k : \operatorname{dom}(C) \subseteq H_0 \to H_k, \quad k \in \{1, \ldots, n\},
$$

are densely defined and linear but not necessarily closable, although C is closed by assumption. We denote by $\iota_{\operatorname{dom}(C) \hookrightarrow H_0}$ the canonical embedding from dom(C) endowed with the graph inner product of C into H_0. Note that $C_k \iota_{\operatorname{dom}(C) \hookrightarrow H_0}$ is a continuous linear operator from both the Hilbert space dom(C) and H_0 (with dense range) into H_k. With regards to the Gelfand triple

$$
\operatorname{dom}(C) \subseteq H_0 \subseteq \operatorname{dom}(C)',
$$

we now have, in suggestive matrix notation, the (continuous) dual operator

$$C^\diamond = \left(C_1^\diamond \cdots C_n^\diamond \right) : H_1 \oplus \cdots \oplus H_n \to \mathrm{dom}\,(C)'$$

$$\begin{pmatrix} v_1 \\ \vdots \\ v_n \end{pmatrix} \mapsto C^\diamond \begin{pmatrix} v_1 \\ \vdots \\ v_n \end{pmatrix}$$

$$\equiv \left(C_1^\diamond \cdots C_n^\diamond \right) \begin{pmatrix} v_1 \\ \vdots \\ v_n \end{pmatrix} \equiv \sum_{k=1}^{n} C_k^\diamond v_k.$$

Thus, in the sense of the embedding $H_0 \subseteq \mathrm{dom}(C)'$

$$\mathrm{dom}\left(C^*\right) = \left\{ \begin{pmatrix} v_1 \\ \vdots \\ v_n \end{pmatrix} \in H_1 \oplus \cdots \oplus H_n \,\middle|\, \sum_{k=1}^{n} C_k^\diamond v_k \in H_0 \right\}.$$

This conceptual framework opens a variety of applications, compare [69] and in particular (2.4.5). We shall also refer to Proposition 2.4.3 for a result of a similar nature.

Requisites from Functional Analysis

B

The material presented here can be found in many books on functional analysis, for example the standard references [11,23,73]. Unfortunately it is not always in the particular flavor and with the particular focus needed for the understanding of our approach. We therefore summarize here a collection of results needed for the constructions presented in this text.

B.1 Fundamentals of Hilbert Space Theory

We start with the basic definitions of binary relations, correspondences, functions and mappings.

Definition B.1.1 Let X and Y be sets.

(1) A subset $a \subseteq X \times Y$ is called a *relation between X and Y*.
(2) If a is a relation between X and Y, we call the pair $(a, X \times Y)$ a *(binary) correspondence*, and we refer to a as the *graph of the correspondence*.
(3) If $(a, X \times Y)$ is a correspondence, we define by

$$a[M] := \{y \in Y \mid \text{there exists } x \in M \text{ such that } (x, y) \in a\}$$

the *post-set of* $M \subseteq X$ and by

$$[N]a := \{x \in X \mid \text{there exists } y \in N \text{ such that } (x, y) \in a\}$$

the *pre-set of* $N \subseteq Y$.

© Springer Nature Switzerland AG 2020

R. Picard et al., *A Primer for a Secret Shortcut to PDEs of Mathematical Physics*, Frontiers in Mathematics, https://doi.org/10.1007/978-3-030-47333-4

(4) A relation a between X and Y is called a *function*, if it is *right-unique*, that is, for each $x \in X$, $y, z \in Y$ we have that

$$(x, y) \in a \text{ and } (x, z) \in a \Rightarrow y = z.$$

(5) If a is a function between X and Y, we call the correspondence $(a, X \times Y)$ a *mapping*.

We emphasize that there is a subtle difference between a relation and a correspondence as well as between a function and a mapping. Indeed, in general it is not possible to reconstruct the (entirety of the) sets X and Y from a relation a. Take for instance the relation

$$a := \left\{ (x, x^2) \mid x \in \mathbb{R} \right\},$$

which is even a function. However, there are several correspondences having a as its graph, for instance,

$$\left(a, \mathbb{R} \times \mathbb{R}_{\geq 0} \right), \ (a, \mathbb{R} \times \mathbb{R}), \ (a, \mathbb{C} \times \mathbb{R}), \ (a, \mathbb{C} \times \mathbb{C}), \text{ etc.}$$

Note that in each case, the function a stays the same but the corresponding mappings differ. For instance, in the first case the mapping turns out to be onto, while the other mappings fail to have this property. This illustrates, that most of the properties, which are frequently attributed to functions are indeed properties of mappings.

We continue by fixing some notation for functions and mappings.

Definition B.1.2 Let X and Y be sets and $(f, X \times Y)$ a mapping. We set

$$\mathrm{dom}(f) := [Y]f$$

and

$$\mathrm{ran}(f) := f[X],$$

the *domain* and the *range of the mapping,* respectively. Moreover, since $f[\{x\}]$ for $x \in \mathrm{dom}(f)$ is a singleton, we denote its element by $f(x)$ and call it the *image of x under f.* To indicate a mapping $(f, X \times Y)$ with its domain $\mathrm{dom}(f)$ and its graph f, we also write

$$f : \mathrm{dom}(f) \subseteq X \to Y$$

$$x \mapsto f(x).$$

If $\text{dom}(f) = X$ we also write

$$f : X \to Y$$

$$x \mapsto f(x).$$

Although, commonly done, it may cause confusion that a mapping and its function are labeled by the same name f. We caution the reader to be aware of this general habit, which we shall also follow.

We now come to the definition of Hilbert spaces. To do so, we need to define what we mean by a linear space.

Definition B.1.3 Let X be a set and $\mathbb{K}$ be a field. Moreover, let

$$+ : X \times X \to X$$

be a mapping, such that $(X, +)$ is an Abelian group. Furthermore, for all $\alpha \in \mathbb{K}$ there is a mapping

$$(\alpha \cdot) : X \to X$$

satisfying the following properties:

(1) For all $\alpha \in \mathbb{K}$, $x, y \in X$

$$(\alpha \cdot)(x + y) = (\alpha \cdot)\,(x) + (\alpha \cdot)\,(y).$$

(2) For all $\alpha, \beta \in \mathbb{K}$, $x \in X$

$$((\alpha + \beta) \cdot)\,(x) = (\alpha \cdot)\,(x) + (\beta \cdot)(y) \text{ and } (\beta \cdot)\,((\alpha \cdot)\,(x)) = (\beta \alpha \cdot)\,(x).$$

(3) For all $x \in X$

$$(1 \cdot)(x) = x,$$

where $1 \in \mathbb{K}$ denotes the unit with respect to multiplication in $\mathbb{K}$.

Then the triple $(X, +, (\alpha \cdot)_{\alpha \in \mathbb{K}})$ is called a *linear space over* $\mathbb{K}$. In particular, if $\mathbb{K} = \mathbb{R}$ or $\mathbb{K} = \mathbb{C}$, we call $(X, +, (\alpha \cdot)_{\alpha \in \mathbb{K}})$ a *real* or *complex linear space,* respectively.

Again, similar to the identification of mappings and functions, one usually labels a linear space $(X, +, (\alpha\cdot)_{\alpha\in\mathbb{K}})$ simply by X and refers to X as a linear space over $\mathbb{K}$. Having the notion of a linear space at hand, we can define inner product spaces and Hilbert spaces.

Definition B.1.4 Let $\mathbb{K} \in \{\mathbb{R}, \mathbb{C}\}$ and X a linear space over $\mathbb{K}$. A mapping

$$\langle\cdot|\cdot\rangle_X : X \times X \to \mathbb{K}$$

$$(x, y) \mapsto \langle x|y\rangle_X$$

is called an *inner product on X,* if

(1) For all $x \in X$ the mapping

$$\langle x|\cdot\rangle_X : X \to \mathbb{K}$$

$$y \mapsto \langle x|y\rangle_X$$

is linear, that is, for all $y, z \in X$ and $\alpha \in \mathbb{K}$ we have that

$$\langle x|\alpha \cdot y + z\rangle_X = \alpha\langle x|y\rangle_X + \langle x|z\rangle_X.$$

(2) For all $x, y \in X$

$$\langle x|y\rangle_X = \langle y|x\rangle_X^*,$$

where α^* denotes the complex conjugate of a complex number $\alpha \in \mathbb{C}$.
(3) For all $x \in X$

$$\langle x|x\rangle_X \in \mathbb{R}_{\geq 0}.$$

(4) For all $x \in X$

$$\langle x|x\rangle_X = 0 \Rightarrow x = 0.$$

If $\langle\cdot|\cdot\rangle_X$ is an inner product on X, we call the pair $(X, \langle\cdot|\cdot\rangle_X)$ (or simply X) an *inner product space.*

As it turns out, inner product spaces are special 'normed' spaces as the next lemma shows.

Lemma B.1.5 *Let $(X, \langle \cdot | \cdot \rangle_X)$ be an inner product space. We define the mapping*

$$| \cdot |_X : X \to \mathbb{R}_{\geq 0}$$

$$x \mapsto \sqrt{\langle x | x \rangle_X}.$$

Then the Cauchy–Schwarz inequality holds: For all $x, y \in X$

$$|\langle x | y \rangle_X| \leq |x|_X |y|_X.$$

Moreover, the mapping $| \cdot |_X$ is a norm on X, that is,

(1) for all $x, y \in X$ the triangle inequality $|x + y|_X \leq |x|_X + |y|_X$ holds,
(2) for all $x \in X$ and $\alpha \in \mathbb{K}$,[1] $|\alpha x|_X = |\alpha|_\mathbb{K} |x|_X$,
(3) for all $x \in X$, $|x|_X = 0 \Rightarrow x = 0$.

Proof Let $x, y \in X$. We observe that due to the properties of an inner product the matrix

$$G_{x,y} := \begin{pmatrix} \langle x | x \rangle_X & \langle x | y \rangle_X \\ \langle y | x \rangle_X & \langle y | y \rangle_X \end{pmatrix} \in \mathbb{K}^{2 \times 2}$$

is an accretive, Hermitian matrix, that is, $G_{x,y}^* = G_{x,y}$ and

$$
\begin{aligned}
z^* G_{x,y} z &= \begin{pmatrix} z_1^* & z_2^* \end{pmatrix} \begin{pmatrix} \langle x | x \rangle_X & \langle x | y \rangle_X \\ \langle y | x \rangle_X & \langle y | y \rangle_X \end{pmatrix} \begin{pmatrix} z_1 \\ z_2 \end{pmatrix} \\
&= \begin{pmatrix} z_1^* & z_2^* \end{pmatrix} \begin{pmatrix} \langle x | x \rangle_X z_1 + \langle x | y \rangle_X z_2 \\ \langle y | x \rangle_X z_1 + \langle y | y \rangle_X z_2 \end{pmatrix} \\
&= \langle x | x \rangle_X z_1^* z_1 + \langle x | y \rangle_X z_1^* z_2 + \langle y | x \rangle_X z_2^* z_1 + \langle y | y \rangle_X z_2^* z_2 \\
&= \langle z_1 x + z_2 y | z_1 x + z_2 y \rangle_X \geq 0
\end{aligned}
$$

for each $z = (z_1, z_2) \in \mathbb{K}^2$. Thus, its determinant is non-negative

$$\langle x | x \rangle_X \langle y | y \rangle_X - \langle x | y \rangle_X \langle y | x \rangle_X \geq 0.$$

Using that $\langle y | x \rangle_X = \langle x | y \rangle_X^*$, we get

$$|\langle x | y \rangle_X|^2 \leq \langle x | x \rangle_X \langle y | y \rangle_X = |x|_X^2 |y|_X^2,$$

---

[1]For simplicity, we shall occasionally dispense with the $\cdot$, when we denote the application of $\alpha \cdot$.

which in turn implies the Cauchy–Schwarz inequality. To prove that $|\cdot|_X$ is a norm, we observe that the properties (b) and (c) immediately follow from the properties of an inner product. To show the triangle inequality, let $x, y \in X$ and compute

$$
\begin{aligned}
|x + y|_X^2 &= \langle x + y | x + y \rangle_X \\
&= |x|_X^2 + 2 \operatorname{\mathfrak{Re}}\langle x | y \rangle_X + |y|_X^2 \\
&\leq |x|_X^2 + 2|x|_X |y|_X + |y|_X^2 \\
&= (|x|_X + |y|_X)^2,
\end{aligned}
$$

where we have used the Cauchy–Schwarz inequality as well as $\operatorname{\mathfrak{Re}}\langle x | y \rangle_X \leq |\langle x | y \rangle_X|$. $\square$

This lemma provides a natural topology for inner product spaces induced by the corresponding norm. Thus, notions like convergence, Cauchy sequence or completeness can be used in the framework of inner product spaces.

Definition B.1.6 Let $(X, \langle \cdot | \cdot \rangle_X)$ be an inner product space. If X is complete with respect to the topology induced by the norm $|\cdot|_X$ given in Lemma B.1.5, we call $(X, \langle \cdot | \cdot \rangle_X)$ (or simply X) a *Hilbert space*. If $\mathbb{K} = \mathbb{R}$, we call X a *real Hilbert space,* and if $\mathbb{K} = \mathbb{C}$, we call X a *complex Hilbert space.*

We shall use the letter H with or without indices to generically designate Hilbert spaces.

Lemma B.1.7 *Let $(H, \langle \cdot | \cdot \rangle_H)$ be an inner product space. Then the* parallelogram identity *holds: for all $x, y \in H$*

$$
|x + y|_H^2 + |x - y|_H^2 = 2 \left(|x|_H^2 + |y|_H^2 \right).
$$

The proof is left as an exercise for the interested reader. It is remarkable that the converse statement of Lemma B.1.7 is also true: Let $(X, |\cdot|_X)$ be a normed space and assume that the parallelogram identity holds. Then, $|\cdot|_X$ is induced by an inner product. The proof of this uses the so-called *polarization equality.*

We now provide a way to construct a new Hilbert space out of a finite number of given Hilbert spaces. We leave the proof of the following proposition as an easy exercise to the reader.

Proposition B.1.8 *Let $n \in \mathbb{N}$ and $H_0, \ldots, H_n$ be Hilbert spaces over the same field $\mathbb{K} \in \{\mathbb{R}, \mathbb{C}\}$. Set $H := H_0 \times \ldots \times H_n$. We equip H with the following operations:*

$$
(x_0, \ldots, x_n) + (y_0, \ldots, y_n) := (x_0 + y_0, \ldots, x_n + y_n)
$$

for all $x_i, y_i \in H_i, i \in \{0, \ldots n\}$ as well as

$$(\alpha \cdot)\,((x_0, \ldots, x_n)) := (\alpha \cdot x_0, \ldots, \alpha \cdot x_n)$$

for all $x_i \in H_i, i \in \{0, \ldots, n\}$ and $\alpha \in \mathbb{K}$. Then $(H, +, (\alpha \cdot)_{\alpha \in \mathbb{K}})$ is a linear space over $\mathbb{K}$. Moreover, setting

$$\langle (x_0, \ldots, x_n) | (y_0, \ldots, y_n) \rangle_H := \langle x_0 | y_0 \rangle_{H_0} + \ldots + \langle x_n | y_n \rangle_{H_n}$$

for all $x_i, y_i \in H_i, i \in \{0, \ldots n\}$, we obtain an inner product on H and $(H, \langle \cdot | \cdot \rangle_H)$ becomes a Hilbert space, called the direct sum of $H_0, \ldots, H_n$ *and denoted by*

$$H_0 \oplus \ldots \oplus H_n.$$

Remark B.1.9 Frequently, we denote the elements of $H_0 \oplus \ldots \oplus H_n$ by column vectors, that is, as vectors of the form

$$\begin{pmatrix} x_0 \\ \vdots \\ x_n \end{pmatrix}$$

where $x_i \in H_i, i \in \{0, \ldots, n\}$.

We conclude this section be explaining the realification of a complex Hilbert space and the complexification of a real Hilbert space.

Proposition B.1.10

(1) Let H be a complex Hilbert space. Then, by restricting the scalar field to $\mathbb{R}$, H becomes a real Hilbert space, $H_\mathbb{R}$, the realification of H, with respect to the inner product

$$\langle x | y \rangle_{H_\mathbb{R}} := \mathfrak{Re}\langle x | y \rangle_H,$$

for all $x, y \in H$.

(2) Let H be a real Hilbert space. Then, $H \times H$ equipped with the element-wise addition and scalar multiplication

$$(\alpha + \mathrm{i}\beta)\,(x, y) := (\alpha x - \beta y, \alpha y + \beta x)$$

for $\alpha, \beta \in \mathbb{R}$ and $x, y \in H$, becomes a complex linear space. Defining

$$\langle (x, y)|(u, v)\rangle_{H_{\mathbb{C}}} := \langle x|u\rangle_H + \langle y|v\rangle_H + i(\langle x|v\rangle_H - \langle y|u\rangle_H)$$

for all $x, y, u, v \in H$, $(H \times H, \langle \cdot | \cdot \rangle_{H_{\mathbb{C}}})$ becomes a complex Hilbert space, called the complexification of H and denoted by $H_{\mathbb{C}}$.

Proof We just prove the second part, since the first one is straight-forward. So, let H be a real Hilbert space. Then it is easy to see that $H \times H$ is a complex linear space with respect to the defined operations. We prove that $\langle \cdot | \cdot \rangle_{H_{\mathbb{C}}}$ is an inner product. Let $x, y, u, v, w, z \in H$ and $\alpha, \beta \in \mathbb{R}$. Then

$\langle (x, y)|(\alpha + i\beta)(u, v) + (w, z)\rangle_{H_{\mathbb{C}}}$

$= \langle (x, y)|(\alpha u - \beta v + w, \alpha v + \beta u + z)\rangle_{H_{\mathbb{C}}}$

$= \langle x|\alpha u - \beta v + w\rangle_H + \langle y|\alpha v + \beta u + z\rangle_H + i(\langle x|\alpha v + \beta u + z\rangle_H - \langle y|\alpha u - \beta v + w\rangle_H)$

$= (\alpha + i\beta)\langle x|u\rangle_H + (i\alpha - \beta)\langle x|v\rangle_H + (\alpha + i\beta)\langle y|v\rangle_H - (i\alpha - \beta)\langle y|u\rangle_H +$

$\quad + \langle x|w\rangle_H + \langle y|z\rangle_H + i\langle x|z\rangle_H - i\langle y|w\rangle_H$

$= (\alpha + i\beta)(\langle x|u\rangle_H + \langle y|v\rangle_H + i(\langle x|v\rangle_H - \langle y|u\rangle_H)) + \langle (x, y)|(w, z)\rangle_{H_{\mathbb{C}}}$

$= (\alpha + i\beta)\langle (x, y)|(u, v)\rangle_{H_{\mathbb{C}}} + \langle (x, y)|(w, z)\rangle_{H_{\mathbb{C}}},$

which shows the linearity in the second argument. Moreover,

$$\langle (x, y)|(u, v)\rangle_{H_{\mathbb{C}}} = \langle x|u\rangle_H + \langle y|v\rangle_H + i(\langle x|v\rangle_H - \langle y|u\rangle_H)$$
$$= \langle u|x\rangle_H + \langle v|y\rangle_H - i(\langle u|y\rangle_H - \langle v|x\rangle_H)$$
$$= \langle (u, v)|(x, y)\rangle_{H_{\mathbb{C}}}^*.$$

Finally, we observe that

$$\langle (x, y)|(x, y)\rangle_{H_{\mathbb{C}}} = \langle x|x\rangle_H + \langle y|y\rangle_H = \langle (x, y)|(x, y)\rangle_{H \oplus H}$$

and thus, the remaining properties of an inner product and the completeness of $H_{\mathbb{C}}$ follow from Proposition B.1.8.   □

B.2   The Projection Theorem

The central theorem of Hilbert space theory is the projection theorem, a variant of which we are formulating next. For doing so, we need to introduce the following notions.

Definition B.2.1 Let H be a Hilbert space. Two elements x, $y \in H$ are called *orthogonal*, denoted by $x \perp y$, if

$$\langle x | y \rangle_H = 0.$$

Moreover, for a subset $M \subseteq H$ we define

$$M^\perp := \{ y \in H \mid \text{for all } x \in M : x \perp y \},$$

the *orthogonal complement* of M.

Remark B.2.2 If x, $y \in H$ are orthogonal, then a direct computation shows

$$|x + y|_H^2 = |x|_H^2 + |y|_H^2. \tag{B.2.1}$$

Moreover, for $M \subseteq H$, the set $M^\perp$ is a closed subspace of H. Indeed, the linearity of the inner product in the second argument yields that $M^\perp$ is a subspace and the Cauchy–Schwarz inequality yields the closedness.

Theorem B.2.3 (Projection Theorem)   *Let H be a Hilbert space and M a closed subspace. Then there exists an onto, continuous, linear mapping $\kappa : H \to M$ such that*

$$|\kappa x - x|_H = \inf \{ |y - x|_H \mid y \in M \}.$$

Moreover,

$$y = \kappa x \Leftrightarrow y \in M, \quad x - y \in M^\perp. \tag{B.2.2}$$

Proof Let $x \in H$. First, we prove that there exists a unique element $\kappa x \in M$ with

$$|\kappa x - x|_H = \inf \{ |y - x|_H \mid y \in M \} =: d.$$

Let $(y_n)_{n \in \mathbb{N}}$ be a sequence in M with

$$|y_n - x|_H \to d \quad (n \to \infty).$$

For $y, z \in M$ we compute, by using the parallelogram identity (see Lemma B.1.7),

$$|x - y|_H^2 + |x - z|_H^2 = |x - \frac{1}{2}(y + z) - \frac{1}{2}(y - z)|_H^2 + |x - \frac{1}{2}(y + z) + \frac{1}{2}(y - z)|_H^2$$

$$= 2 \left(|x - \frac{1}{2}(y + z)|_H^2 + |\frac{1}{2}(y - z)|_H^2 \right)$$

$$\geq 2d^2 + \frac{1}{2}|y - z|_H^2,$$

where we have used that $\frac{1}{2}(y + z) \in M$, since M is a linear subspace. Thus,

$$\frac{1}{2}|y - z|_H^2 \leq |x - y|_H^2 + |x - z|_H^2 - 2d^2 \tag{B.2.3}$$

for each $y, z \in M$. In particular, choosing $y = y_n$ and $z = y_m$ for $m, n \in \mathbb{N}$ we derive

$$\frac{1}{2}|y_n - y_m|_H^2 \leq |x - y_n|_H^2 + |x - y_m|_H^2 - 2d^2 \to 0 \quad (n, m \to \infty),$$

that is, $(y_n)_{n \in \mathbb{N}}$ is a Cauchy-sequence. Thus, it is convergent and we denote its limit by y_0. Since M is closed, we deduce that $y_0 \in M$ and derive

$$|y_0 - x|_H = \lim_{n \to \infty} |y_n - x|_H = d.$$

Moreover, if $\tilde{y} \in M$ satisfies

$$|\tilde{y} - x|_H = d,$$

then (B.2.3) gives

$$\frac{1}{2}|y_0 - \tilde{y}|_H^2 \leq |x - y_0|_H^2 + |x - \tilde{y}|_H^2 - 2d^2 = 0$$

and so, $\tilde{y} = y_0$. Thus, we have proved that there exists a mapping $\kappa : H \to M$, $x \mapsto \kappa x$ with

$$|\kappa x - x|_H = \inf \{|y - x|_H \mid y \in M\} \quad (x \in H).$$

Obviously, $\kappa x = x$ if $x \in M$ and thus, κ is onto. Before we prove the remaining properties of κ, we show the characterization (B.2.2). We first prove that $x - \kappa x \in M^\perp$. Let $y \in M$. Then, for $t \in \mathbb{K}$ we compute

$$d^2 \le |x - \kappa x - ty|_H^2$$
$$= |x - \kappa x|_H^2 - 2\,\Re\langle x - \kappa x | ty \rangle_H + t^2 |y|_H^2$$
$$= d^2 - 2\,\Re\, t \langle x - \kappa x | y \rangle_H + t^2 |y|_H^2$$

so that

$$2\,\Re\, t \langle x - \kappa x | y \rangle_H \le t^2 |y|_H^2.$$

Choosing $t = \pm\frac{1}{n}$ and $t = \pm i\frac{1}{n}$, (for $\mathbb{K} = \mathbb{R}$, we can dispend with the latter choice of t) and letting n tend to infinity, we derive

$$\langle x - \kappa x | y \rangle_H = 0,$$

that is, $x - \kappa x \in M^\perp$. Now, let $y \in M$ such that $x - y \in M^\perp$. Then, for each $z \in M$,

$$|x - z|_H^2 = |x - y + (y - z)|_H^2 = |x - y|_H^2 + |y - z|_H^2 \ge |x - y|_H^2,$$

where we have used (B.2.1). Thus,

$$|x - y|_H = \inf\{|x - z|_H \mid z \in M\}$$

and hence, from above $y = \kappa x$.

It is left to show that κ is linear and continuous. Let $x, \tilde{x} \in H$ and $\lambda \in \mathbb{K}$. By (B.2.2) $\kappa(\lambda x + \tilde{x}) = \lambda \kappa x + \kappa \tilde{x}$ is equivalent to

$$\lambda x + \tilde{x} - (\lambda \kappa x + \kappa \tilde{x}) \in M^\perp.$$

This, however, is clear since $x - \kappa x, \tilde{x} - \kappa \tilde{x} \in M^\perp$ and $M^\perp$ is a subspace. Finally, for $x \in H$ we estimate, using (B.2.1)

$$|x|_H^2 = |x - \kappa x + \kappa x|_H^2 = |x - \kappa x|_H^2 + |\kappa x|_H^2 \ge |\kappa x|_H^2,$$

which proves the continuity of κ.   □

As a consequence of Theorem B.2.3 we have the following decomposition result.

Corollary B.2.4 (Orthogonal Decomposition) *Let H be a Hilbert space and M a closed subspace. Then*

$$U : H \to M \oplus M^\perp$$

$$x \mapsto \begin{pmatrix} \kappa x \\ x - \kappa x \end{pmatrix}$$

is an isometric linear surjection with inverse

$$U^{-1} : M \oplus M^\perp \to H$$

$$\begin{pmatrix} u \\ v \end{pmatrix} \mapsto u + v.$$

Proof By Theorem B.2.3 the mapping κ is linear and thus, so is U. Moreover, U is onto. Indeed, for $u \in M$ and $v \in M^\perp$ we set $x := u + v$. Then by (B.2.2), $u = \kappa x$, since $x - u = v \in M^\perp$ and consequently $x - \kappa x = x - u = v$, which shows $Ux = (u, v)$, that is, U is onto. The isometry of U follows from (B.2.1), since

$$|Ux|^2_{M \oplus M^\perp} = |\kappa x|^2_H + |x - \kappa x|^2_H = |x|^2_H$$

for each $x \in H$. This completes the proof. $\square$

Isometric, linear surjections between Hilbert spaces are called *unitary* mappings. The close relationship between H and $M \oplus M^\perp$ allows to identify both and so one usually writes briefly

$$H = M \oplus M^\perp \tag{B.2.4}$$

and speaks of an *orthogonal sum*.

Remark B.2.5 If $\iota_{M \hookrightarrow H} : M \to H$ denotes the canonical isometric, linear embedding of M in H, that is, $\iota_{M \hookrightarrow H} x = x$ then

$$P_M := \iota_{M \hookrightarrow H} \kappa : H \to H$$

is a mapping known as the *orthogonal projector onto* M. Orthogonal projectors are continuous linear operators P characterized by the properties:

$$PP = P \text{ and } P[H] \perp (1 - P)[H]. \tag{B.2.5}$$

Corollary B.2.6 *Let H be a Hilbert space and $M \subseteq H$. Then*

$$M^{\perp\perp} = \overline{\operatorname{span} M}.$$

In particular, a subspace $U \subseteq H$ is dense in H if and only if $U^{\perp} = \{0\}$.

Proof It is clear that $M \subseteq M^{\perp\perp}$ and since $M^{\perp\perp}$ is a closed subspace, we deduce

$$\overline{\operatorname{span} M} \subseteq M^{\perp\perp}.$$

Moreover, $\left(\overline{\operatorname{span} M}\right)^{\perp} = M^{\perp}$. Indeed, the inclusion $\left(\overline{\operatorname{span} M}\right)^{\perp} \subseteq M^{\perp}$ is clear. Let $x \in M^{\perp}$, that is,

$$M \subseteq [\{0\}]\langle x|\cdot\rangle_H.$$

Since $\langle x|\cdot\rangle_H : H \to \mathbb{K}$ is a linear and continuous mapping, its kernel $[\{0\}]\langle x|\cdot\rangle_H$ is a closed subspace and hence,

$$\overline{\operatorname{span} M} \subseteq [\{0\}]\langle x|\cdot\rangle_H,$$

and thus, $x \in \left(\overline{\operatorname{span} M}\right)^{\perp}$. Now let $x \in M^{\perp\perp}$. By Corollary (B.2.4) we can decompose $x = y + z$ with $y \in \overline{\operatorname{span} M}$ and $z \in \left(\overline{\operatorname{span} M}\right)^{\perp} = M^{\perp}$. Using that $y \in M^{\perp\perp}$, we see that $z = x - y \in M^{\perp\perp}$ and hence, $z = 0$. Thus $x = y \in \overline{\operatorname{span} M}$, which proves the equality

$$M^{\perp\perp} = \overline{\operatorname{span} M}.$$

To prove the last claim, let $U \subseteq H$ be a subspace. Then U is dense, if and only if $U^{\perp\perp} = \overline{U} = H$. Hence, if $U^{\perp} = \{0\}$, the subspace is dense and conversely, if U is dense we obtain $U^{\perp} = U^{\perp\perp\perp} = H^{\perp} = \{0\}$. □

B.3   The Riesz Representation Theorem

The dual space H' of a Hilbert space is given by the set

$$\{f : H \to \mathbb{K} \mid f \text{ continuous and linear}\}$$

equipped with the linear structure

$$(f + \alpha \cdot g)\, x := fx + \alpha^* gx \tag{B.3.1}$$

for $f, g \in H'$ and $\alpha \in \mathbb{K}$. Note the conjugation of α in this definition, which is a harmless modification to adapt to the conjugate linearity of the inner product in the first factor. H' equipped with the norm

$$|f|_{H'} := \sup |\{fx \,|\, x \in B\,(0,1)\}|,$$

where $B\,(0, 1)$ denotes the closed ball around 0 with radius 1 in H, is complete, that is, a Banach space. The Riesz representation theorem now yields that H' is actually a Hilbert space.

Theorem B.3.1 (Riesz' Representation Theorem)   *Let H be a Hilbert space. Then for every $f \in H'$ there is a unique $x_f \in H$ such that*

$$fz = \langle x_f | z \rangle_H \;\text{ for all } z \in H.$$

Proof Let $f \in H'$. We start by proving the uniqueness of such a representing element. Assume that $x, y \in H$ represent f. Then, for each $z \in H$ we have

$$\langle x - y | z \rangle_H = \langle x | z \rangle_H - \langle y | z \rangle_H = 0$$

and thus, $x = y$. Now we prove the existence of such an element. Since f is linear and continuous, the kernel of f

$$N := [\{0\}]f$$

is a closed linear subspace of H. By Corollary B.2.4 we have

$$H = N \oplus N^\perp.$$

If $N^\perp = \{0\}$, we have that $N = H$ and thus, $fz = 0$ for each $z \in H$. Hence, we may choose $x_f = 0$. If $N^\perp \neq \{0\}$, we find an element $x \in N^\perp$ with $|x|_H = 1$. Since $x \notin N$, we have $fx \neq 0$. Thus, for each $z \in H$,

$$z - \frac{fz}{fx}x \in N.$$

Therefore,

$$0 = \langle x | z - \frac{fz}{fx}x \rangle_H = \langle x | z \rangle_H - \frac{fz}{fx}|x|_H^2 = \langle x | z \rangle_H - \frac{fz}{fx}$$

or equivalently

$$fz = \langle (fx)^* x | z \rangle_H$$

for each $z \in H$.   □

Proposition B.3.2 *Let H be a Hilbert space. We define*

$$R_H : H' \to H$$

$$f \mapsto x_f,$$

where $x_f \in H$ is chosen according to Theorem B.3.1. Then R_H is an isometric, linear surjection, called the Riesz *mapping of H.*

Proof Let $f, g \in H', \alpha \in \mathbb{K}$. Then for each $z \in H$ we have that

$$(\alpha f + g)z = \alpha^* fz + gz$$

$$= \alpha^* \langle R_H f | z \rangle_H + \langle R_H g | z \rangle_H$$

$$= \langle \alpha R_H f + R_H g | z \rangle_H,$$

which shows $R_H(\alpha f + g) = \alpha R_H f + R_H g$, i.e. R_H is linear. Moreover,

$$|R_H f|_H^2 = \langle R_H f | R_H f \rangle_H = f(R_H f) \leq |f|_{H'} |R_H f|_H,$$

which proves

$$|R_H f|_H \leq |f|_{H'}$$

for all $f \in H'$. Moreover, for $f \in H'$ and $z \in H$, $|z|_H \leq 1$, we estimate

$$|fz| = |\langle R_H f | z \rangle_H| \leq |R_H f|_H |z|_H \leq |R_H f|_H$$

and thus,

$$|f|_{H'} \leq |R_H f|_H,$$

which proves the isometry. Finally, for $x \in H$, the mapping $\langle x | \cdot \rangle_H : H \to \mathbb{K}$ is linear and continuous, that is, $\langle x | \cdot \rangle_H \in H'$. Then clearly,

$$R_H \langle x | \cdot \rangle_H = x,$$

which shows that R_H is onto.   □

Corollary B.3.3 *Let H be a Hilbert space. Then H' is a Hilbert space with inner product*

$$\langle f | g \rangle_{H'} := \langle R_H f | R_H g \rangle_H$$

for $f, g \in H'$.

Remark B.3.4 In the Hilbert space case, one usually identifies H with H'' (the dual of H'), so that

$$\varphi f = (f \varphi)^*$$

for $\varphi \in H'' = H, f \in H'$, and we shall follow this convention. Then the Riesz mapping $R_{H'} : H'' = H \to H'$ is actually the inverse of R_H. Indeed, using the isometry of R_H we get

$$\langle R_H R_{H'} \varphi | x \rangle_H = \left(R_{H'} \varphi | R_H^{-1} x \right)_{H'} = \varphi \left(R_H^{-1} x \right) = \left(\left(R_H^{-1} x \right) \varphi \right)^* = \langle x | \varphi \rangle_H^* = \langle \varphi | x \rangle_H$$

for all $x \in H, \varphi \in H'' = H$, from which we read off that

$$R_H R_{H'} = 1$$

or

$$R_{H'} = R_H^{-1}.$$

B.4   Linear Operators and Their Adjoints

Throughout, let H_0, H_1 be Hilbert spaces over the same field $\mathbb{K} \in \{\mathbb{R}, \mathbb{C}\}$.

Definition B.4.1 Let

$$A : \mathrm{dom}\,(A) \subseteq H_0 \to H_1,$$

be a linear mapping (also called *linear operator*)

(1) A is called *densely defined*, if dom(A) is dense in H_0, that is, $\overline{\mathrm{dom}(A)} = H_0$.
(2) A is called *closed*, if A considered as a subset of $H_0 \oplus H_1$ is closed. That is, for each sequence $(x_n)_{n \in \mathbb{N}}$ in dom(A) with $x_n \to x$ and $A x_n \to y$ as $n \to \infty$ for some $x \in H_0, y \in H_1$, it follows that $x \in \mathrm{dom}(A)$ and $y = Ax$.
(3) A is called *closable*, if $\overline{A}$ is a function, where the closure is taken in $H_0 \oplus H_1$.

Lemma B.4.2 *Let $A : \mathrm{dom}(A) \subseteq H_0 \to H_1$ be a linear operator.*

(1) The mapping

$$\langle \cdot | \cdot \rangle_{\mathrm{dom}(A)} : \mathrm{dom}(A) \times \mathrm{dom}(A) \to \mathbb{K}$$

$$(x, y) \mapsto \langle x | y \rangle_{H_0} + \langle Ax | Ay \rangle_{H_1}$$

is an inner product on $\mathrm{dom}(A)$, the graph inner product. Moreover, $(\mathrm{dom}(A), \langle \cdot | \cdot \rangle_{\mathrm{dom}(A)})$ is a Hilbert space if and only if A is closed.

(2) A is closable if and only if for each sequence $(x_n)_{n \in \mathbb{N}}$ with $x_n \to 0$ and $Ax_n \to y$ as $n \to \infty$ for some $y \in H_1$ it follows that $y = 0$.

Proof

(1) Observe that for $x, y \in \mathrm{dom}(A)$ we have that

$$\langle x | y \rangle_{\mathrm{dom}(A)} = \langle (x, Ax) | (y, Ay) \rangle_{H_0 \oplus H_1}. \tag{B.4.1}$$

Hence, the properties of an inner product for $\langle \cdot | \cdot \rangle_{\mathrm{dom}(A)}$ follow immediately from Proposition B.1.8. Moreover, if A is closed, then $(A, \langle \cdot | \cdot \rangle_{H_0 \oplus H_1})$ is a Hilbert space as a closed subspace of $H_0 \oplus H_1$. Thus, $(\mathrm{dom}(A), \langle \cdot | \cdot \rangle_{\mathrm{dom}(A)})$ is complete. Conversely, if $\mathrm{dom}(A)$ is a Hilbert space then by (B.4.1), $(A, \langle \cdot | \cdot \rangle_{H_0 \oplus H_1})$ is a Hilbert space and thus, A is a closed subspace of $H_0 \oplus H_1$.

(2) Let A be closable. If $(x_n)_{n \in \mathbb{N}}$ is a sequence in $\mathrm{dom}(A)$ with $x_n \to 0$ and $Ax_n \to y$ as $n \to \infty$ for some $y \in H_1$, then $(0, y) \in \overline{A}$, that is, $0 = \overline{A}0 = y$. To show the converse, let $(x, y), (x, z) \in \overline{A}$ for some $x \in H_0, y, z \in H_1$. We have to show that $y = z$. Since A is linear, so is $\overline{A}$ and thus,

$$(0, y - z) \in \overline{A}.$$

Hence, there is a sequence $(x_n)_{n \in \mathbb{N}}$ in $\mathrm{dom}(A)$ such that $x_n \to 0$ and $Ax_n \to y - z$ as $n \to \infty$. By assumption, it follows that $y = z$. □

Before we come to the definition of adjoint operators, we briefly recall some well-known facts about continuous linear operators. We recall the closed unit ball $B_{H_0}(0, 1) := \{x \in H_0 \mid |x|_{H_0} \leq 1\}$ (occasionally, we will omit the subscript on $B_{H_0}(0, 1)$ if the Hilbert space is clear from the context).

Lemma B.4.3 *Denote by $\mathcal{B}(H_0, H_1)$ the set of all linear continuous operators $L : H_0 \to H_1$. Moreover, for $L \in \mathcal{B}(H_0, H_1)$ we define the norm of L as*

$$\|L\| := \sup \left\{ |Lx|_{H_1} \mid x \in B_{H_0}(0, 1) \right\}.$$

Then $(\mathcal{B}(H_0, H_1), \|\cdot\|)$ is a Banach space.

Proof Obviously, $\mathcal{B}(H_0, H_1)$ is a vector space with the usual point-wise addition and scalar multiplication. We first prove that $\|L\| < \infty$ for $L \in \mathcal{B}(H_0, H_1)$. Indeed, since L is continuous, it is in particular continuous at 0. Thus, there is $\delta > 0$ such that

$$|x|_{H_0} \le \delta \Rightarrow |Lx|_{H_1} < 1$$

for each $x \in H_0$. Hence, for $x \in B_{H_0}(0, 1)$ we estimate

$$|Lx|_{H_1} = \frac{1}{\delta} |L(\delta x)|_{H_1} < \frac{1}{\delta},$$

which shows $\|L\| < \infty$. The properties of a norm for $\|\cdot\|$ are easy to verify. Thus, it is left to prove that $(\mathcal{B}(H_0, H_1), \|\cdot\|)$ is complete. Let $(L_n)_{n\in\mathbb{N}}$ be a Cauchy-sequence in $\mathcal{B}(H_0, H_1)$. Then, for each $x \in H_0, n, m \in \mathbb{N}$,

$$|L_n x - L_m x|_{H_1} \le \|L_n - L_m\| |x|_{H_0},$$

which shows that $(L_n x)_{n\in\mathbb{N}}$ is a Cauchy-sequence in H_1 and thus, convergent. We set

$$Lx := \lim_{n\to\infty} L_n x.$$

It is easy to see that L is linear. Given $\varepsilon > 0$ choose $N \in \mathbb{N}$ such that for all $n, m \ge N$

$$\|L_n - L_m\| < \varepsilon.$$

Then for all $n, m \ge N$, we have for all $x \in B(0, 1)$

$$|L_n x - L_m x|_H < \varepsilon.$$

Hence, letting $m \to \infty$ we deduce that for all $n \ge N$ and $x \in B(0, 1)$ we have

$$|L_n x - Lx|_{H_1} \le \varepsilon.$$

or, equivalently, for all $n \geq N$

$$\sup_{x \in B(0,1)} |L_n x - L x|_{H_1} \leq \varepsilon \tag{B.4.2}$$

In particular, the latter gives that $L - L_N$ is continuous. Indeed, for $x, y \in H_0$ with $x \neq y$ we see that

$$| (L - L_N) x - (L - L_N) y|_{H_1} = |x - y|_{H_0} \left| (L - L_N) \left(\frac{x - y}{|x - y|_{H_0}} \right) \right|_{H_1} \leq \varepsilon |x - y|_{H_0},$$

proving that $L - L_N$ is Lipschitz-continuous. Thus, $L = L - L_N + L_N$ is continuous, that is, $L \in \mathcal{B}(H_0, H_1)$. Moreover, (B.4.2) shows that L is indeed the limit of $(L_n)_{n \in \mathbb{N}}$ with respect to $\| \cdot \|$.   □

Definition B.4.4 Let $A : \mathrm{dom}(A) \subseteq H_0 \to H_1$ be a linear operator. We define the *adjoint relation* $A^* \subseteq H_1 \times H_0$ *of* A by

$$(x, y) \in A^* \Leftrightarrow \text{ for all } u \in \mathrm{dom}(A) : \langle Au | x \rangle_{H_1} = \langle u | y \rangle_{H_0}.$$

Lemma B.4.5 *Let* $A : \mathrm{dom}(A) \subseteq H_0 \to H_1$ *be a linear operator. Then* A^* *is a closed subspace of* $H_1 \oplus H_0$. *Moreover,* A^* *is a function, if and only if* A *is densely defined. In this case, we call* A^* *the* adjoint operator *of* A.

Proof Since

$$(x, y) \in A^* \Leftrightarrow \text{ for all } u \in \mathrm{dom}(A) : \langle (u, Au), (y, -x) \rangle_{H_0 \oplus H_1} = 0 \Leftrightarrow (y, -x) \in A^{\perp}, \tag{B.4.3}$$

we easily derive that A^* is a closed subspace from the corresponding properties of orthogonal complements. Now, due to linearity, A^* is a function if and only if $A^*[\{0\}] = \{0\}$. However, we have that

$$y \in A^*[\{0\}] \Leftrightarrow (0, y) \in A^*$$

$$\Leftrightarrow \text{ for all } u \in \mathrm{dom}(A) : \langle u | y \rangle_{H_0} = \langle Au | 0 \rangle_{H_1} = 0$$

$$\Leftrightarrow y \in \mathrm{dom}(A)^{\perp}.$$

Hence, $A^*[\{0\}] = \mathrm{dom}(A)^{\perp}$. Thus, A^* is a function if and only if $\mathrm{dom}(A)^{\perp} = \{0\}$, which is equivalent to the density of $\mathrm{dom}(A)$ by Corollary B.2.6.   □

Lemma B.4.6 *Let $A : \text{dom}(A) \subseteq H_0 \to H_1$ be a linear operator, which is injective. Then $\left(A^{-1}\right)^* = (A^*)^{-1}$.*

Proof Let $(x, y) \in H_0 \times H_1$. Then

$$(x, y) \in \left(A^*\right)^{-1} \Leftrightarrow (y, x) \in A^*$$

$$\Leftrightarrow \text{for all } u \in \text{dom}(A) : \langle Au|y\rangle_{H_1} = \langle u|x\rangle_{H_0}$$

$$\Leftrightarrow \text{for all } v \in A[H_0] = \text{dom}(A^{-1}) : \langle v|y\rangle_{H_1} = \langle A^{-1}v|x\rangle_{H_0}$$

$$\Leftrightarrow (x, y) \in \left(A^{-1}\right)^*$$

and thus, $(A^*)^{-1} = \left(A^{-1}\right)^*$. $\square$

Corollary B.4.7 *Let $A : \text{dom}(A) \subseteq H_0 \to H_1$ be a densely defined linear operator. Then A is closable if and only if A^* is densely defined and in this case we have*

$$\overline{A} = A^{**}.$$

Proof Using (B.4.3), we observe that

$$(u, v) \in A^{\perp\perp} \Leftrightarrow \text{for all } (x, y) \in A^\perp : \langle (x, y)|(u, v)\rangle_{H_0 \oplus H_1} = 0$$

$$\Leftrightarrow \text{for all } x \in \text{dom}(A^*) : \langle (-A^*x, x)|(u, v)\rangle_{H_0 \oplus H_1} = 0$$

$$\Leftrightarrow \text{for all } x \in \text{dom}(A^*) : \langle A^*x|u\rangle_{H_0} = \langle x|v\rangle_{H_1}$$

$$\Leftrightarrow (u, v) \in A^{**}.$$

Hence, $A^{**} = A^{\perp\perp}$. Now, by Corollary B.2.6, $A^{\perp\perp} = \overline{A}$ and thus, A is closable if and only if A^{**} is a function, which in turn is equivalent to A^* being densely defined by Lemma B.4.5. $\square$

Theorem B.4.8 (Projection Theorem, Variant 2) *Let $A : \text{dom}(A) \subseteq H_0 \to H_1$ be a densely defined, closed linear operator. Then we have the orthogonal decompositions*

$$H_1 = \overline{A[H_0]} \oplus [\{0\}] A^*,$$

$$H_0 = \overline{A^*[H_1]} \oplus [\{0\}] A.$$

Proof For the first decomposition it suffices to prove $\overline{(A[H_0])}^{\perp} = [\{0\}]A^*$, by Corollary B.2.4. We have that

$$y \in \overline{(A[H_0])}^{\perp} = (A[H_0])^{\perp} \Leftrightarrow \text{for all } x \in \text{dom}(A) : \langle Ax|y\rangle_{H_1} = 0 = \langle x|0\rangle_{H_0}$$

$$\Leftrightarrow (y, 0) \in A^*$$

$$\Leftrightarrow y \in [\{0\}]A^*.$$

The second decomposition follows by replacing A by A^* (note that due to the closedness of A, $A^{**} = A$ by Corollary B.4.7). □

The somewhat subtle domain issues encountered in connection with general densely defined, linear mappings and their adjoints disappear for left-total, continuous linear operators, that is, continuous linear operators defined on all of H_0.

Proposition B.4.9 *Let $L : H_0 \to H_1$ be a continuous liner operator. Then, $L^* : H_1 \to H_0$ is also continuous and linear with $\|L\| = \|L^*\|$.* .

Proof We first show that $\text{dom}(L^*) = H_1$. Let $y \in H_1$ and consider the mapping

$$\varphi_y : H_0 \to \mathbb{K}$$

$$x \mapsto \langle y|Lx\rangle_{H_1}.$$

Obviously, φ_y is linear and by the Cauchy–Schwarz inequality we estimate

$$|\varphi_y(x)| = |\langle y|Lx\rangle_{H_1}| \le |y|_{H_1}|Lx|_{H_1} \le |y|_{H_1}\|L\||x|_{H_0}$$

for all $x \in H_0$, which proves that φ_y is continuous. In other words, $\varphi_y \in H_0'$ and thus,

$$\langle y|Lx\rangle_{H_1} = \varphi_y(x) = \langle R_{H_0}\varphi_y|x\rangle_{H_0}$$

for all $x \in H_0$. Consequently, $y \in \text{dom}(L^*)$ with $L^*y = R_{H_0}\varphi_y$. It is left to prove the continuity of L^* and the asserted norm equality. By what we have shown so far, we have that

$$|L^*y|_{H_0} = |R_{H_0}\varphi_y|_{H_0} = |\varphi_y|_{H_0'} \le \|L\||y|_{H_1},$$

which shows the continuity of L^* with $\|L^*\| \le \|L\|$. Moreover, for $x \in H_0$ we estimate

$$|Lx|^2_{H_1} = \langle Lx|Lx\rangle_{H_1} = \langle L^*Lx|x\rangle_{H_0} \le |L^*Lx|_{H_0}|x|_{H_0} \le \|L^*\||Lx|_{H_1}|x|_{H_0},$$

which shows

$$|Lx|_{H_1} \leq \|L^*\| |x|_{H_0}$$

and thus, the norm equality follows.   □

Proposition B.4.10 *Let $A : \mathrm{dom}(A) \subseteq H_0 \to H_1$ be a densely defined linear operator and X a Hilbert space over the same field as H_0 and H_1.*

(1) Let $L : H_0 \to H_1$ be linear and continuous. Then $(A + L)^* = A^* + L^*$.
(2) Let $L : H_1 \to X$ be linear and continuous. Then $(LA)^* = A^* L^*$.
(3) Let $B : \mathrm{dom}(B) \subseteq X \to H_0$ be densely defined, bijective on $\mathrm{dom}(B)$ and linear, such that $B^{-1} : H_0 \to X$ is continuous. Then AB is densely defined and $(AB)^* = B^* A^*$.

Proof

(1) Let $y \in \mathrm{dom}(A^*) = \mathrm{dom}(A^*) \cap \mathrm{dom}(L^*) = \mathrm{dom}(A^* + L^*)$ by Proposition B.4.9. Then for $x \in \mathrm{dom}(A)$

$$\begin{aligned}
\langle (A + L)x | y \rangle_{H_1} &= \langle Ax | y \rangle_{H_1} + \langle Lx | y \rangle_{H_1} \\
&= \langle x | A^* y \rangle_{H_0} + \langle x | L^* y \rangle_{H_0} \\
&= \langle x | (A^* + L^*) y \rangle_{H_0},
\end{aligned}$$

which proves $y \in \mathrm{dom}((A + L)^*)$ with $(A + L)^* y = (A^* + L^*) y$. Thus, we have shown

$$A^* + L^* \subseteq (A + L)^*.$$

Let now $y \in \mathrm{dom}((A + L)^*)$. Then

$$\begin{aligned}
\langle Ax | y \rangle_{H_1} &= \langle (A + L) x | y \rangle_{H_1} - \langle Lx | y \rangle_{H_1} \\
&= \langle x | (A + L)^* y \rangle_{H_0} - \langle x | L^* y \rangle_{H_0} \\
&= \langle x | ((A + L)^* - L^*) y \rangle_{H_0}
\end{aligned}$$

for all $x \in \mathrm{dom}(A)$, and hence, $y \in \mathrm{dom}(A^*) = \mathrm{dom}(A^* + L^*)$ with $A^* y = ((A + L)^* - L^*) y$, which gives $(A + L)^* y = A^* y + L^* y$, showing the reverse inclusion.
(2) Let $y \in \mathrm{dom}(A^* L^*)$, that is, $L^* y \in \mathrm{dom}(A^*)$. Then

$$\langle LAx | y \rangle_X = \langle Ax | L^* y \rangle_{H_1} = \langle x | A^* L^* y \rangle_{H_0}$$

for all $x \in \mathrm{dom}(A) = \mathrm{dom}(LA)$, which shows $A^*L^* \subseteq (LA)^*$. If $y \in \mathrm{dom}\,((LA)^*)$, we obtain

$$\langle Ax|L^*y\rangle_{H_1} = \langle LAx|y\rangle_X = \langle x|(LA)^*y\rangle_{H_0},$$

for all $x \in \mathrm{dom}(A)$, which gives $L^*y \in \mathrm{dom}(A^*)$ and $(LA)^* \subseteq A^*L^*$.

(3) Let $y \in \mathrm{dom}(B^*A^*)$, that is, $y \in \mathrm{dom}(A^*)$ and $A^*y \in \mathrm{dom}(B^*)$. Then we have

$$\langle ABx|y\rangle_{H_1} = \langle Bx|A^*y\rangle_{H_0} = \langle x|B^*A^*y\rangle_X$$

for all $x \in \mathrm{dom}(AB)$ and thus, $B^*A^* \subseteq (AB)^*$. Conversely, let $(y, z) \in (AB)^*$. Then,

$$\langle Ax|y\rangle_{H_1} = \langle ABB^{-1}x|y\rangle_{H_1} = \langle B^{-1}x|z\rangle_X = \langle x|\left(B^{-1}\right)^*z\rangle_{H_0}$$

for all $x \in \mathrm{dom}(A)$ and thus, $y \in \mathrm{dom}(A^*)$ with $A^*y = (B^{-1})^*z$. Using $\left(B^{-1}\right)^* = (B^*)^{-1}$ (Lemma B.4.6), we derive $A^*y \in \mathrm{dom}(B^*)$ with $B^*A^*y = z$ and thus, $(AB)^* \subseteq B^*A^*$. Since B^*A^* is a function, we see that AB is densely defined according to Lemma B.4.5.   □

Definition B.4.11 Let $A : \mathrm{dom}(A) \subseteq H \to H$ be a densely defined linear operator in a Hilbert space H.

(1) A is called *symmetric*, if $A \subseteq A^*$, that is, $\mathrm{dom}(A) \subseteq \mathrm{dom}(A^*)$ and $A^*x = Ax$ for all $x \in \mathrm{dom}(A)$.
(2) A is called *skew-symmetric,* if $A \subseteq -A^*$, that is, $\mathrm{dom}(A) \subseteq \mathrm{dom}(A^*)$ and $-A^*x = Ax$ for all $x \in \mathrm{dom}(A)$.
(3) A is called *selfadjoint*, if $A = A^*$.
(4) A is called *skew-selfadjoint,* if $A = -A^*$.

Remark B.4.12 Note that (skew-)selfadjoint operators are automatically closed by Lemma B.4.5. Moreover, in case of a complex Hilbert space, A is symmetric/selfadjoint if and only if iA is skew-symmetric/skew-selfadjoint.

Whereas (skew-)symmetry is easily established (skew-)selfadjointness is a subtle property. The following result gives a useful criterion.

Theorem B.4.13 *Let $A : \mathrm{dom}\,(A) \subseteq H \to H$ be a closed skew-symmetric operator on a Hilbert space H. Then the following statements are equivalent*

(1) A is skew-selfadjoint,
(2) for all $\lambda \in \mathbb{C} \setminus i\mathbb{R}$ the operator $\lambda + A$ is onto,
(3) there is $\lambda \in \mathbb{C}$ such that $\lambda + A$ and $\lambda^* - A$ are onto.

Moreover, for skew-selfadjoint operators we have that $\lambda + A$ is injective and $(\lambda + A)^{-1}$ is continuous for each $\lambda \in \mathbb{C} \setminus i\mathbb{R}$.

Proof (1) $\Rightarrow$ (2). We set $\lambda = \eta + i\mu$ with $\eta, \mu \in \mathbb{R}$ and $\eta \neq 0$. For $x \in \mathrm{dom}(A)$ we compute

$$\mathfrak{Re}\langle (\lambda^* - A) x | x \rangle_H = \eta |x|_H^2 - \mathfrak{Re}\langle Ax|x\rangle_H.$$

Since A is skew-selfadjoint,

$$\langle Ax|x\rangle_H = \langle x| - Ax\rangle_H = -\langle Ax|x\rangle_H^*,$$

and thus,

$$\mathfrak{Re}\langle Ax|x\rangle_H = 0.$$

Thus, we can estimate

$$|\eta||x|_H^2 = \left| \mathfrak{Re}\langle (\lambda^* - A) x|x\rangle_H \right| \leq |(\lambda^* - A)x|_H |x|_H.$$

In particular, we see that $\lambda^* - A$ is injective, since $\eta \neq 0$. Thus, by Theorem B.4.8 and Proposition B.4.10 we have that

$$\left(\overline{(\lambda + A)[H]} \right)^{\perp} = [\{0\}] (\lambda + A)^* = [\{0\}](\lambda^* - A) = \{0\}.$$

Hence, $\lambda + A$ has dense range, by Corollary B.2.6. To show the claim we need to prove that $(\lambda + A)[H]$ is closed. So, let $(x_n)_{n \in \mathbb{N}}$ be a sequence in $\mathrm{dom}(A)$ with $(\lambda + A) x_n \to y$ for some $y \in H$ as $n \to \infty$. As above, with λ^* replaced by λ and $-A$ replaced by A, we estimate

$$|\eta||x_n - x_m|_H^2 \leq |(\lambda + A)(x_n - x_m)|_H |x_n - x_m|_H$$

for all $n, m \in \mathbb{N}$, which shows that $(x_n)_{n \in \mathbb{N}}$ is a Cauchy-sequence in H and thus convergent. Setting $x := \lim_{n \to \infty} x_n$, we get $Ax_n = (\lambda + A) x_n - \lambda x_n \to y - \lambda x$ as $n \to \infty$ and hence, by the closedness of A, $x \in \mathrm{dom}(A)$ with $Ax = y - \lambda x$, which shows $y \in (\lambda + A)[H]$. Thus, the range of $\lambda + A$ is dense and closed, hence, $\lambda + A$ is onto.

(2) $\Rightarrow$ (3). We choose $\lambda = 1$. By assumption $1 + A$ and $-1 + A$ are onto, which yields the assertion.

(3) $\Rightarrow$ (1). Let $\lambda \in \mathbb{C}$ such that $\lambda + A$ and $\lambda^* - A$ are onto. Since A is skew-symmetric, it suffices to prove $A^* \subseteq -A$. Let $y \in \mathrm{dom}(A^*)$. Then there exists $x \in \mathrm{dom}(A)$ such that

$$(\lambda + A) x = (\lambda - A^*) y.$$

We claim that $x = y$. Indeed, note that, by skew-symmetry, $x \in \text{dom}(A^*)$ and $(\lambda + A) x = (\lambda - A^*)x$. Thus, if we can prove that $\lambda - A^*$ is injective, the assertion follows. Using Theorem B.4.8 we have that

$$[\{0\}](\lambda - A^*) = \left((\lambda^* - A)[H]\right)^{\perp} = \{0\},$$

since $\lambda^* - A$ is onto. Thus, $y = x \in \text{dom}(A)$ which shows that $A = -A^*$.

The last assertion follows from the fact that

$$|(\lambda - A) x|_H |x|_H \geq |\mathfrak{Re}\langle (\lambda - A) x | x \rangle_H| = |\mathfrak{Re}\,\lambda| |x|_H^2$$

for $x \in \text{dom}(A)$, and thus $(\lambda - A)^{-1}$ is continuous, if $\lambda \in \mathbb{C} \backslash i\mathbb{R}$ with $\|(\lambda - A)^{-1}\| \leq \frac{1}{|\mathfrak{Re}\,\lambda|}$. Since $\lambda - A$ is also onto by (2), the claim follows. $\qquad\square$

For the symmetric case the argument is analogous.

Corollary B.4.14 *Let $A : \text{dom}(A) \subseteq H \to H$ be a closed symmetric operator on a Hilbert space H, and there is a $\lambda \in \mathbb{C}$ such that $\lambda + A$ and $\lambda^* + A$ are onto. Then A is selfadjoint.*

Proof The result follows from Theorem B.4.13 with A replaced by iA and λ by $i\lambda$. $\qquad\square$

We now discuss important examples of selfadjoint and skew-selfadjoint operators. We begin with the following easy lemma.

Lemma B.4.15 *Let $A : \text{dom}(A) \subseteq H_0 \to H_1$ and $B : \text{dom}(B) \subseteq H_1 \to H_0$ be two densely defined linear operators and define*

$$\begin{pmatrix} 0 & B \\ A & 0 \end{pmatrix} : \text{dom}(A) \times \text{dom}(B) \subseteq H_0 \oplus H_1 \to H_0 \oplus H_1$$

$$(x, y) \mapsto (By, Ax).$$

Then

$$\begin{pmatrix} 0 & B \\ A & 0 \end{pmatrix}^* = \begin{pmatrix} 0 & A^* \\ B^* & 0 \end{pmatrix}.$$

Proof Let $(u, v) \in \text{dom}(B^*) \times \text{dom}(A^*)$. Then we compute for each $(x, y) \in \text{dom}(A) \times \text{dom}(B)$:

$$\left\langle \begin{pmatrix} 0 & B \\ A & 0 \end{pmatrix} \begin{pmatrix} x \\ y \end{pmatrix} \middle| \begin{pmatrix} u \\ v \end{pmatrix} \right\rangle_{H_0 \oplus H_1} = \langle By|u \rangle_{H_0} + \langle Ax|v \rangle_{H_1}$$

$$= \langle y|B^*u \rangle_{H_1} + \langle x|A^*v \rangle_{H_0}$$

$$= \left\langle \begin{pmatrix} x \\ y \end{pmatrix} \middle| \begin{pmatrix} 0 & A^* \\ B^* & 0 \end{pmatrix} \begin{pmatrix} u \\ v \end{pmatrix} \right\rangle_{H_0 \oplus H_1},$$

which shows

$$\begin{pmatrix} 0 & A^* \\ B^* & 0 \end{pmatrix} \subseteq \begin{pmatrix} 0 & B \\ A & 0 \end{pmatrix}^*.$$

On the other hand, if $(u, v) \in \text{dom}\left(\begin{pmatrix} 0 & B \\ A & 0 \end{pmatrix}^* \right)$, for all $x \in \text{dom}(A)$

$$\langle Ax|v \rangle_{H_1} = \left\langle \begin{pmatrix} 0 & B \\ A & 0 \end{pmatrix} \begin{pmatrix} x \\ 0 \end{pmatrix} \middle| \begin{pmatrix} u \\ v \end{pmatrix} \right\rangle_{H_0 \oplus H_1}$$

$$= \left\langle \begin{pmatrix} x \\ 0 \end{pmatrix} \middle| \begin{pmatrix} 0 & A \\ B & 0 \end{pmatrix}^* \begin{pmatrix} u \\ v \end{pmatrix} \right\rangle_{H_0 \oplus H_1}$$

$$= \langle x|f \rangle_{H_0},$$

where $f \in H_0$ denotes the first component of $\begin{pmatrix} 0 & A \\ B & 0 \end{pmatrix}^* \begin{pmatrix} u \\ v \end{pmatrix}$. Thus, $v \in \text{dom}(A^*)$.
Similarly, we can show $u \in \text{dom}(B^*)$ and thus, the assertion follows.   □

Corollary B.4.16 *Let* $A : \text{dom}(A) \subseteq H_0 \to H_1$ *be a densely defined closed linear operator. Then* $\begin{pmatrix} 0 & -A^* \\ A & 0 \end{pmatrix}$ *is skew-selfadjoint.*

Proof Note that A^* is densely defined by Corollary B.4.7. Thus, by Lemma B.4.15

$$\begin{pmatrix} 0 & -A^* \\ A & 0 \end{pmatrix}^* = \begin{pmatrix} 0 & A^* \\ -A^{**} & 0 \end{pmatrix} = - \begin{pmatrix} 0 & -A^* \\ A & 0 \end{pmatrix},$$

due to the closedness of A.   □

The following result often used as a criterion to establish selfadjointness. We add the remarkably elementary proof.

Proposition B.4.17 *Let* $A : D(A) \subseteq H_0 \to H_1$ *be a densely defined closed linear operator. Then* A^*A *and* AA^* *are selfadjoint.*

Proof Consider

$$B := \begin{pmatrix} 0 & -A^* \\ A & 0 \end{pmatrix},$$

which is skew-selfadjoint by Corollary B.4.16. Hence, by Theorem B.4.13, the operator $1 - B$ is bijective and $(1 - B)^{-1}$ is continuous. Thus, by Proposition B.4.10 (3)

$$((1 + B)(1 - B))^* = (1 - B)^*(1 + B)^* = (1 + B)(1 - B),$$

which shows that $1 - B^2 = (1 + B)(1 - B)$ is selfadjoint. Since

$$1 - B^2 = \begin{pmatrix} 1 + A^*A & 0 \\ 0 & 1 + AA^* \end{pmatrix},$$

we deduce that A^*A and AA^* are densely defined. Moreover,

$$
\begin{aligned}
\begin{pmatrix} 1 + A^*A & 0 \\ 0 & 1 + AA^* \end{pmatrix} &= \begin{pmatrix} 1 + A^*A & 0 \\ 0 & 1 + AA^* \end{pmatrix}^* \\
&= \left(\begin{pmatrix} 0 & 1 \\ 1 & 0 \end{pmatrix} \begin{pmatrix} 0 & 1 + AA^* \\ 1 + A^*A & 0 \end{pmatrix} \right)^* \\
&= \begin{pmatrix} 0 & 1 + AA^* \\ 1 + A^*A & 0 \end{pmatrix}^* \begin{pmatrix} 0 & 1 \\ 1 & 0 \end{pmatrix}^* \\
&= \begin{pmatrix} 0 & (1 + A^*A)^* \\ (1 + AA^*)^* & 0 \end{pmatrix} \begin{pmatrix} 0 & 1 \\ 1 & 0 \end{pmatrix} \\
&= \begin{pmatrix} (1 + A^*A)^* & 0 \\ 0 & (1 + AA^*)^* \end{pmatrix},
\end{aligned}
$$

where we have used Proposition B.4.10 (3) and Lemma B.4.15. The latter gives $1 + A^*A = (1 + A^*A)^* = 1 + (A^*A)^*$ and $1 + AA^* = (1 + AA^*)^* = 1 + (AA^*)^*$, which shows the assertion. $\qquad\square$

The concept of adjoints can be used to illuminate the structure of orthogonal projectors.

Lemma B.4.18 (Orthogonal Projectors) *Let $M \subseteq H$ be a closed subspace of a Hilbert space H and let*

$$\iota_{M \hookrightarrow H} : M \to H$$

denote the canonical embedding of M in H. Then the orthogonal projector P_M of H onto M can be factorized as

$$P_M = \iota_{M \hookrightarrow H} \iota_{M \hookrightarrow H}^*.$$

Thus, the mapping $\kappa : H \to M$ in Theorem B.2.3 is nothing but $\iota_{M \hookrightarrow H}^*$. In particular, P_M is selfadjoint.

Proof Let $x \in H$. Since $P_M x$ can be characterized as the element $z \in M$ satisfying

$$\langle z | y \rangle_H = \langle x | y \rangle_H$$

for all $y \in M$, by Theorem B.2.3 we have that

$$
\begin{aligned}
\langle P_M x | y \rangle_H &= \langle x | \iota_{M \hookrightarrow H} y \rangle_H \\
&= \langle \iota_{M \hookrightarrow H}^* x | y \rangle_M \\
&= \langle \iota_{M \hookrightarrow H} \iota_{M \hookrightarrow H}^* x | \iota_{M \hookrightarrow H} y \rangle_H
\end{aligned}
$$

and so,

$$P_M x = \iota_{M \hookrightarrow H} \iota_{M \hookrightarrow H}^* x. \qquad \square$$

B.5   Duals and Adjoints

We now consider the construction of a dual operator. Throughout, let H_0, H_1 be Hilbert spaces over the same field $\mathbb{K} \in \{\mathbb{R}, \mathbb{C}\}$.

Definition B.5.1 Let $A \in \mathcal{B}(H_0, H_1)$. Then we define the *dual operator* $A' : H_1' \to H_0'$ by

$$\left(A' \varphi \right)(x) := \varphi(Ax)$$

for each $x \in H_0$, $\varphi \in H_1'$.

Remark B.5.2 Obviously, if $\varphi \in H_1'$ then $A'\varphi$ is a linear mapping form H_0 to $\mathbb{K}$ and

$$|(A'\varphi)(x)| \leq |\varphi|_{H_1'} \|A\| |x|_{H_0}$$

which shows that $A'\varphi$ is indeed an element of H_0'. Furthermore, A' itself is linear and

$$\|A'\| = \sup \left\{ |A'\varphi|_{H_0'} \mid \varphi \in H_1', |\varphi|_{H_1'} \leq 1 \right\}$$

$$\leq \|A\|$$

by what we have shown above. Thus, $A' \in \mathcal{B}(H_1', H_0')$ with $\|A'\| \leq \|A\|$. Moreover for $x \in H_0$ we set $\varphi := R_{H_1}^{-1}(Ax) = \langle Ax|\cdot\rangle_{H_1} \in H_1'$ and estimate

$$|Ax|_{H_1}^2 = \left|\langle Ax|Ax\rangle_{H_1}\right| = |\varphi(Ax)| = \left|(A'\varphi)(x)\right| \leq \|A'\| |\varphi|_{H_1'} |x|_{H_0} = \|A'\| |Ax|_{H_1} |x|_{H_0},$$

which yields $|Ax|_{H_1} \leq \|A'\| |x|_{H_0}$, showing that $\|A\| \leq \|A'\|$. Thus, we obtain

$$\|A\| = \|A'\|.$$

There is a strong connection between the concepts of dual and adjoint operators.

Proposition B.5.3 *Let $A \in \mathcal{B}(H_0, H_1)$. Then $A^* = R_{H_0} A' R_{H_1}^{-1}$.*

Proof Let $x \in H_0$, $y \in H_1$. We need to prove that

$$\langle y|Ax\rangle_{H_1} = \langle R_{H_0} A' R_{H_1}^{-1} y|x\rangle_{H_0}.$$

Indeed, we have that

$$\langle R_{H_0} A' R_{H_1}^{-1} y|x\rangle_{H_0} = \left(A' R_{H_1}^{-1} y\right)(x)$$

$$= \left(R_{H_1}^{-1} y\right)(Ax)$$

$$= \langle y|Ax\rangle_{H_1},$$

which shows the claim.   □

Also the adjoints of non-continuous operators can be expressed with the help of dual operators, see also Proposition 2.4.3. In order to do so, we introduce so-called Gelfand-triples.

Definition Let X_0 be a Hilbert space over $\mathbb{K}$ with $X_0 \subseteq H_0$ such that

$$\iota_{X_0 \hookrightarrow H_0} : X_0 \to H_0$$

$$x \mapsto x$$

is continuous and has dense range. Then we call (X_0, H_0, X_0') a *Gelfand triple*.

Lemma B.5.4 *Let (X_0, H_0, X_0') be a Gelfand triple. Then $H_0' \subseteq X_0'$ via*

$$\left(\iota_{X_0 \hookrightarrow H_0}\right)' : H_0' \to X_0',$$

which is injective and has dense range. Moreover, $\left(\iota_{X_0 \hookrightarrow H_0}\right)'(\varphi) = \varphi|_{X_0}$ for each $\varphi \in H_0'$.

Proof By Proposition B.5.3 we have $\left(\iota_{X_0 \hookrightarrow H_0}\right)' = R_{H_0}^{-1} \iota_{X_0 \hookrightarrow H_0}^* R_{H_1}$. Thus, it suffices to prove that $\iota_{X_0 \hookrightarrow H_0}^*$ is injective and has dense range. This however follows from Theorem B.4.8, since

$$[\{0\}]\iota_{X_0 \hookrightarrow H_0}^* = \left(\overline{\left(\iota_{X_0 \hookrightarrow H_0}\right)[X_0]}\right)^{\perp} = H_0^{\perp} = \{0\},$$

$$\overline{\iota_{X_0 \hookrightarrow H_0}^*[H_0]} = \left([\{0\}]\iota_{X_0 \hookrightarrow H_0}\right)^{\perp} = \{0\}^{\perp} = X_0.$$

Moreover,

$$\left(\iota_{X_0 \hookrightarrow H_0}'(\varphi)\right)(x) = \varphi(\iota_{X_0 \hookrightarrow H_0} x) = \varphi(x)$$

for all $\varphi \in H_0'$, $x \in X_0$ and thus, the last assertion holds. $\square$

Definition Let (X_0, H_0, X_0') be a Gelfand triple and let $A : X_0 \subseteq H_0 \to H_1$ be a linear operator, such that $A\iota_{X_0 \hookrightarrow H_0} \in \mathcal{B}(X_0, H_1)$. Then we define

$$A^{\diamond} := (A\iota_{X_0 \hookrightarrow H_0})' R_{H_1}^{-1} \in \mathcal{B}(H_1, X_0').$$

Remark B.5.5 This definition can be applied to $\iota : X_0 \subseteq H_0 \to H_0$, $x_0 \mapsto x_0$. Then, for $y \in H_0$, $x \in X_0$ we obtain

$$\left(\iota^{\diamond} y\right)(x) = \left(\iota' R_{H_0}^{-1} y\right)(x) = \langle y | x \rangle_{H_0}.$$

The purpose of constructing this continuous dual operator $A^{\diamond}$ is to generalize A^* by avoiding the delicate domain issues needed for constructing the adjoint A^* of a possibly unbounded linear operator A.

Proposition B.5.6 *Let (X_0, H_0, X_0') be a Gelfand triple and let $A : X_0 \subseteq H_0 \to H_1$ be a linear operator, such that $A\iota_{X_0 \hookrightarrow H_0} \in \mathcal{B}(X_0, H_1)$. Then*

$$\mathrm{dom}(A^*) = [H_0']A^\diamond$$

and

$$A^* y = R_{H_0} A^\diamond y \quad (y \in \mathrm{dom}(A^*)).$$

Proof Let $y \in \mathrm{dom}(A^*) \subseteq H_1$. Then for all $x \in X_0$ we have

$$\langle Ax | y \rangle_{H_1} = \langle x | A^* y \rangle_{H_0}.$$

Consequently,

$$\left(A^\diamond y\right)(x) = \left((A\iota_{X_0 \hookrightarrow H_0})' R_{H_1}^{-1} y\right)(x) = \langle y | Ax \rangle_{H_1} = \langle A^* y | x \rangle_{H_0}$$

for all $x \in X_0$ and therefore,

$$A^\diamond y = \left(\langle A^* y | \cdot \rangle_{H_0}\right)|_{X_0},$$

which clearly can be extended to H_0. Thus, $A^\diamond y \in H_0'$. Conversely, if $y \in H_1$ such that $A^\diamond y \in H_0'$ we have that

$$\langle R_{H_0} A^\diamond y | x \rangle_{H_0} = \left(A^\diamond y\right)(x) = \langle y | Ax \rangle_{H_1}$$

for all $x \in X_0$ and so $y \in \mathrm{dom}(A^*)$ with

$$A^* y = R_{H_0} A^\diamond y,$$

which completes the proof. $\square$

B.6   Solution Theory for (Real) Strictly Positive Linear Operators

Let H be a Hilbert space over $\mathbb{K} \in \{\mathbb{R}, \mathbb{C}\}$ and $A : \mathrm{dom}(A) \subseteq H \to H$ linear.

Definition B.6.1 We define the quadratic form q_A by

$$q_A(x) := \langle x | Ax \rangle_H \quad (x \in D(A)).$$

Moreover, we set

$$w(A) := \overline{q_A[S_H(0,1)]},$$

where $S_H(0, 1)$ denotes the unit sphere, that is, the boundary of the unit ball $B_H(0, 1)$ in H centered at the origin. The set $w(A)$ is called the *numerical range of A*. The operator A is called *strictly accretive* or *strictly positive definite* (or just *strictly positive*), if

$$\inf \mathfrak{Re}[w(A)] \geq c_0,$$

for some $c_0 > 0$. A is called *accretive*, if

$$\inf \mathfrak{Re}[w(A)] \geq 0,$$

It turns out that strict accretivity of an operator and its adjoint leads to a comprehensive solution theory for many applied problems (see in particular Sect. 1.3)

Lemma B.6.2 *Let $A : \mathrm{dom}(A) \subseteq H \to H$ be a linear operator. If A is strictly accretive, then*

$$A^{-1} : A[H] \subseteq H \to H$$

is a well-defined, continuous linear operator.

Proof By assumption, there is $c_0 > 0$ such that

$$c_0 \langle x|x \rangle_H \leq \mathfrak{Re} \langle x|Ax \rangle_H \leq |x|_H |Ax|_H$$

for all $x \in \mathrm{dom}(A)$. From this we read off that A is injective and

$$\left| A^{-1}(Ax) \right|_H = |x|_H \leq \frac{1}{c_0} |Ax|_H,$$

that is,

$$\left\| A^{-1} \right\| = \sup \left\{ \left| A^{-1} y \right|_H \mid y \in A[H], \, |y|_H \leq 1 \right\} \leq \frac{1}{c_0},$$

which shows the continuity of A^{-1}. $\square$

Proposition B.6.3 *Let* $A : \mathrm{dom}\,(A) \subseteq H \to H$ *be a densely defined, closed linear operator. If* A *and* A^* *are strictly accretive, then*

$$A^{-1} : H \to H$$

is a well-defined, continuous linear mapping.

Proof By Lemma B.6.2 we know that

$$A^{-1} : A[H] \subseteq H \to H$$

is a continuous linear mapping. Since

$$H = \overline{A\,[H]} \oplus [\{0\}]A^*$$

by Theorem B.4.8 and $[\{0\}]A^* = \{0\}$ by Lemma B.6.2, we read off that A^{-1} is densely defined. Since A^{-1} is continuous and closed it follows that $A\,[H] = \overline{A\,[H]} = H$.   □

B.7   An Approximation Result

We record the following approximation result, which is needed for our approach to evolutionary equations.

Lemma B.7.1 (Approximation/Regularization) *Let* $A : \mathrm{dom}\,(A) \subseteq H \to H$ *be a densely defined, closed linear operator such that* A *and* A^* *are accretive. Then for each* $x \in H$

$$(1 + \varepsilon A)^{-1}x \to x \quad (\varepsilon \to 0+).$$

Proof We note that $1 + \varepsilon A$ and $(1 + \varepsilon A)^* = 1 + \varepsilon A^*$ are both strictly accretive, and thus $(1 + \varepsilon A)^{-1} \in \mathcal{B}(H)$ for all $\varepsilon > 0$ by Proposition B.6.3. Moreover, the accretivity of εA gives the uniform bound

$$\left\| (1 + \varepsilon A)^{-1} \right\| \leq 1$$

for all $\varepsilon \geq 0$. Since

$$(1 + \varepsilon A)^{-1} - 1 = -\varepsilon A\,(1 + \varepsilon A)^{-1} \tag{B.7.1}$$

we see that for $u \in \mathrm{dom}\,(A)$

$$(1 + \varepsilon A)^{-1} u - u = -\varepsilon A (1 + \varepsilon A)^{-1} u$$
$$= -\varepsilon (1 + \varepsilon A)^{-1} Au$$

and

$$\left| (1 + \varepsilon A)^{-1} Au \right|_H \leq |Au|_H$$

yields

$$(1 + \varepsilon A)^{-1} u - u \overset{\varepsilon \to 0+}{\to} 0.$$

By the density of $\mathrm{dom}\,(A)$ in H and the uniform boundedness noted above, it follows that the latter convergence statement holds for all $u \in H$.   □

It also follows from Lemma B.7.1 that

$$A (1 + \varepsilon A)^{-1} u = (1 + \varepsilon A)^{-1} Au \overset{\varepsilon \to 0+}{\to} Au$$

for all $u \in \mathrm{dom}\,(A)$.

Note that (B.7.1) implies

$$A (1 + \varepsilon A)^{-1} = \varepsilon^{-1} \left(\varepsilon A (1 + \varepsilon A)^{-1} \right)$$
$$= \varepsilon^{-1} \left(1 - (1 + \varepsilon A)^{-1} \right), \tag{B.7.2}$$

which shows that $A (1 + \varepsilon A)^{-1}$ is also bounded[2] in H for $\varepsilon \in \,]0, \infty[$.

B.8   The Root of Selfadjoint Accretive Operators and the Polar Decomposition

This last section is devoted to the polar decomposition of a densely defined closed linear operator $A : \mathrm{dom}(A) \subseteq H_0 \to H_1$ between two Hilbert spaces H_0, H_1 over the same field $\mathbb{K} \in \{\mathbb{R}, \mathbb{C}\}$. Here, we need the concept of a root of a selfadjoint accretive linear operator. We start by studying the case of bounded operators.

---

[2]In the literature the right-hand side of (B.7.2) appears under the name: Yosida approximation (of A).

Lemma B.8.1 *Let H be a Hilbert space and $A \in \mathcal{B}(H)$ be selfadjoint. Then*

$$\|A\| = \sup_{|x|_H=1} |\langle Ax|x\rangle_H|.$$

Proof We set

$$M := \sup_{|x|_H=1} |\langle Ax|x\rangle_H|.$$

Obviously, $M \leq \|A\|$ by the Cauchy–Schwarz inequality. Moreover, we have

$$\langle A(x+y)|x+y\rangle_H - \langle A(x-y)|x-y\rangle_H = 4\,\mathfrak{Re}\langle Ax|y\rangle_H$$

for each $x, y \in H$. Hence,

$$\mathfrak{Re}\langle Ax|y\rangle_H \leq \frac{M}{4}\left(|x+y|_H^2 + |x-y|_H^2\right) = \frac{M}{2}(|x|_H^2 + |y|_H^2)$$

due to Lemma B.1.7. Thus,

$$|Ax|_H = \mathfrak{Re}\langle Ax|\frac{Ax}{|Ax|_H}\rangle_H \leq \frac{M}{2}(|x|_H^2 + 1)$$

and, computing the supremum over $x \in B(0,1)$ on both the left-hand and the right-hand side of this inequality, we obtain

$$\|A\| \leq M,$$

which completes the proof. $\square$

Theorem B.8.2 *Let H be a Hilbert space and $A \in \mathcal{B}(H)$ selfadjoint and accretive. Then there exists a unique selfadjoint and accretive operator $T \in \mathcal{B}(H)$ such that*

$$T^2 = A.$$

The operator T is called the root of A *and is denoted by $\sqrt{A}$. Moreover,*

(1) $[\{0\}]\sqrt{A} = [\{0\}]A$,
(2) if $R \in \mathcal{B}(H)$ and $AR = RA$, then $\sqrt{A}R = R\sqrt{A}$.

Proof We divide the proof into 5 steps. Step 1 and 2 are devoted to the existence of a root, while in Step 3, 4, and 5 we prove its uniqueness and the statements (1) and (2).

Step 1:   We assume

$$\forall x \in H : 0 \le \langle Ax|x \rangle_H \le |x|_H^2 .$$

In particular, we obtain

$$\forall x \in H : \ 0 \le \langle (1 - A)x|x \rangle_H \le |x|_H^2 . \tag{B.8.1}$$

We recall that the Taylor series for the function $[-1, 1] \ni x \mapsto \sqrt{1 - x}$ is given by

$$\sqrt{1 - x} = \sum_{n=0}^{\infty} (-1)^n \binom{\frac{1}{2}}{n} x^n$$

and that this series converges uniformly to $\sqrt{1 - \cdot}$ on $[-1, 1]$. Using (B.8.1), we derive from Lemma B.8.1

$$\| 1 - A \| = \sup_{|x|_H = 1} \langle (1 - A)\, x | x \rangle \le 1,$$

and thus, the series

$$T := \sum_{n=0}^{\infty} (-1)^n \binom{\frac{1}{2}}{n} (1 - A)^n$$

converges in $\mathcal{B}(H)$. Then

$$T^* = \left(\sum_{n=0}^{\infty} (-1)^n \binom{\frac{1}{2}}{n} (1 - A)^n \right)^* = \sum_{n=0}^{\infty} (-1)^n \binom{\frac{1}{2}}{n} \left(1 - A^* \right)^n = T$$

and thus, T is selfadjoint. Next, we show that T is accretive: Since $(-1)^0 \binom{\frac{1}{2}}{0} = 1$ and $(-1)^n \binom{\frac{1}{2}}{n} \le 0$ for $n > 0$, we estimate for all $x \in H$, using $\| 1 - A \| \le 1$,

$$\langle Tx|x \rangle = \left\langle \sum_{n=0}^{\infty} (-1)^n \binom{\frac{1}{2}}{n} (1 - A)^n \, x | x \right\rangle$$

$$= \| x \|^2 + \sum_{n=1}^{\infty} (-1)^n \binom{\frac{1}{2}}{n} \left\langle (1 - A)^n \, x | x \right\rangle$$

$$\geq \|x\|^2 + \sum_{n=1}^{\infty}(-1)^n \binom{\frac{1}{2}}{n}\|(1-A)\|^n\|x\|^2$$

$$\geq \|x\|^2 \sum_{n=0}^{\infty}(-1)^n \binom{\frac{1}{2}}{n} = \|x\|^2\sqrt{1-1} = 0,$$

which confirms that T is accretive. Finally, recalling

$$\binom{\alpha+\beta}{n} = \sum_{k=0}^{n}\binom{\alpha}{k}\binom{\beta}{n-k}$$

for $\alpha, \beta \in \mathbb{R}, n \in \mathbb{N}$, we derive, using the Cauchy product formula,

$$T^2 = \sum_{n=0}^{\infty}\sum_{k=0}^{n}(-1)^k\binom{\frac{1}{2}}{k}(1-A)^k(-1)^{n-k}\binom{\frac{1}{2}}{n-k}(1-A)^{n-k}$$

$$= \sum_{n=0}^{\infty}\left(\sum_{k=0}^{n}\binom{\frac{1}{2}}{k}\binom{\frac{1}{2}}{n-k}\right)(-1)^n(1-A)^n$$

$$= \sum_{n=0}^{\infty}\binom{1}{n}(-1)^n(1-A)^n$$

$$= 1 - (1-A)$$

$$= A,$$

and thus, T is a root of A.

Step 2:   Let $A \in B(H)$ be selfadjoint and accretive. Then $\|A\|^{-1}A$ satisfies (B.8.1) and we set

$$T := \sqrt{\|A\|}\sum_{n=0}^{\infty}(-1)^n\binom{\frac{1}{2}}{n}(1-\|A\|^{-1}A)^n.$$

Clearly, $T \in \mathcal{B}(H)$ is selfadjoint and accretive. Moreover

$$T^2 = \|A\|\|A\|^{-1}A = A$$

and thus, T is a root.

Step 3:   Let T be as in Step 2 and $R \in \mathcal{B}(H)$ with $AR = RA$, then $TR = RT$. Indeed,

$$(1-\|A\|^{-1}A)^n R = R(1-\|A\|^{-1}A)^n$$

and consequently $TR = RT$. This proves (2), once we have shown that T is the unique root of A.

Step 4:   We prove the uniqueness. Let $R \in \mathcal{B}(H)$ be a root of A and T be defined as in Step 2. We show $R = T$. We first note that by Step 3, $TR = RT$. Moreover, since $R + T$ is selfadjoint, we can decompose H as

$$H = [\{0\}](R + T) \oplus \overline{(R + T)[H]},$$

by Theorem B.4.8. For $x \in H$

$$(R - T)(R + T)x = \left(R^2 - T^2 \right) x = (A - A)x = 0$$

and thus, $R = T$ on $(R + T)[H]$ and by continuity on $\overline{(R + T)[H]}$. Thus, we are left with showing the equality on $[\{0\}](R+T)$. Let $x \in [\{0\}](R+T)$. Since R is selfadjoint and accretive, it has a (selfadjoint) root S by Step 2. We have that

$$|Sx|_H^2 = \langle Rx|x \rangle \le \langle (R + T)x|x \rangle_H = 0$$

and hence, $Sx = 0$. However, this implies $x \in [\{0\}]R$, since $Rx = SSx = 0$. Analogously, we obtain $x \in [\{0\}]T$ and thus, $Rx = 0 = Tx$, which shows that $R = T$ in $[\{0\}](R + T)$. Thus, the root of A is uniquely determined.

Step 5:   We prove (1). Obviously, $[\{0\}]\sqrt{A} \subseteq [\{0\}]A$. Assume now that $x \in [\{0\}]A$. Then we have

$$|\sqrt{A}x|_H^2 = \langle \sqrt{A}x|\sqrt{A}x \rangle_H = \langle Ax|x \rangle_H = 0$$

and thus, $\sqrt{A}x = 0$, that is, $x \in [\{0\}]\sqrt{A}$. This proves (1).   □

We now generalize this result to unbounded selfadjoint, accretive operators. We need the following auxiliary result.

Lemma B.8.3 *Let H be a Hilbert space, $B \in \mathcal{B}(H)$ and $C : \mathrm{dom}(C) \subseteq H \to H$ be two selfadjoint, accretive operators. Moreover, assume there exists $\lambda \in \mathbb{C}$ with $(\lambda + C)^{-1} \in \mathcal{B}(H)$ and*

$$(\lambda + C)^{-1}B = B(\lambda + C)^{-1}.$$

Then

$$CB = \overline{BC} = \overline{\sqrt{B}C\sqrt{B}}.$$

Proof By Theorem B.8.2 we have that

$$(\lambda + C)^{-1}\sqrt{B} = \sqrt{B}(\lambda + C)^{-1}.$$

For $x \in \text{dom}(BC) = \text{dom}(C)$,

$$\sqrt{B}x = \sqrt{B}(\lambda + C)^{-1}(\lambda + C)x = (\lambda + C)^{-1}\sqrt{B}(\lambda + C)x$$

and thus, $\sqrt{B}x \in \text{dom}(C)$ with

$$C\sqrt{B}x = (\lambda + C)\sqrt{B}x - \lambda\sqrt{B}x = \sqrt{B}(\lambda + C)x - \lambda\sqrt{B}x = \sqrt{B}Cx$$

and consequently

$$\sqrt{B}C\sqrt{B}x = BCx.$$

Hence,

$$BC \subseteq \sqrt{B}C\sqrt{B}.$$

Moreover, if $x \in \text{dom}\left(\sqrt{B}C\sqrt{B}\right)$, then $\sqrt{B}x \in \text{dom}(C)$ and thus, by what we have shown above, $Bx = \sqrt{B}\sqrt{B}x \in \text{dom}(C)$ with

$$CBx = \sqrt{B}C\sqrt{B}x.$$

Summarizing we have shown

$$BC \subseteq \sqrt{B}C\sqrt{B} \subseteq CB.$$

Since C is closed and B is continuous, we deduce that CB is closed and hence BC and $\sqrt{B}C\sqrt{B}$ are closable with

$$\overline{BC} \subseteq \overline{\sqrt{B}C\sqrt{B}} \subseteq CB.$$

It is left to show $CB = \overline{BC}$. Let $x \in \text{dom}(CB)$. We define $x_\varepsilon := (1 + \varepsilon C)^{-1}x \in \text{dom}(C)$ for $\varepsilon > 0$. Then, by Lemma B.7.1, $x_\varepsilon \to x$ as $\varepsilon \to 0$. Since $BC \subseteq CB$ we have that

$$B(1 + \varepsilon C) \subseteq (1 + \varepsilon C)B$$

and consequently

$$(1 + \varepsilon C)^{-1} B = (1 + \varepsilon C)^{-1} B (1 + \varepsilon C)(1 + \varepsilon C)^{-1} = B(1 + \varepsilon C)^{-1}$$

for each $\varepsilon > 0$. Hence, we obtain

$$BCx_\varepsilon = CBx_\varepsilon = CB(1 + \varepsilon C)^{-1} x = (1 + \varepsilon C)^{-1} CBx \to CBx \quad (\varepsilon \to 0)$$

again by Lemma B.7.1. This shows $x \in \mathrm{dom}(\overline{BC})$ with $\overline{BC}x = CBx$. Hence, $\overline{BC} = CB$.

$\square$

Theorem B.8.4 *Let H be a Hilbert space and $A : \mathrm{dom}(A) \subseteq H \to H$ be selfadjoint and accretive. Then there exists a unique selfadjoint and accretive operator $T : \mathrm{dom}(T) \subseteq H \to H$ such that*

$$T^2 = A.$$

This operator T is called the root of A *and denoted by* $\sqrt{A}$.

Proof We first prove the existence of a root: Consider the operator

$$R := A(1 + A)^{-1} = 1 - (1 + A)^{-1} \in \mathcal{B}(H).$$

This operator is selfadjoint and accretive. Moreover, $(1 + A)^{-1} \in \mathcal{B}(H)$ is selfadjoint and accretive. We set

$$S := \sqrt{(1 + A)^{-1}} \in \mathcal{B}(H)$$

and get

$$[\{0\}]S = [\{0\}](1 + A)^{-1} = (1 + A)[\{0\}] = \{0\},$$

by Theorem B.8.2. Thus, S is injective. By Lemma B.4.6, S^{-1} is selfadjoint and accretive. Define

$$T := S^{-1}\sqrt{R}.$$

Note that $(1 + A)^{-1} R = R(1 + A)^{-1}$ and hence, by Theorem B.8.2, $SR = RS$ and thus, again by Theorem B.8.2,

$$S\sqrt{R} = \sqrt{R}S.$$

Thus, we can apply Lemma B.8.3 (with $\lambda = 0$ and $C = S^{-1}$, $B = \sqrt{R}$) and deduce that

$$T = S^{-1}\sqrt{R} = \overline{\sqrt{R}S^{-1}} = \overline{\sqrt{\sqrt{R}S^{-1}}\sqrt{\sqrt{R}}}.$$

This implies that T is selfadjoint and accretive. Indeed, we have that

$$T^* = \left(\overline{\sqrt{R}S^{-1}}\right)^* = \left(\sqrt{R}S^{-1}\right)^* = S^{-1}\sqrt{R} = T$$

by Proposition B.4.10 and for $x \in \mathrm{dom}\left(\sqrt{\sqrt{R}S^{-1}}\sqrt{\sqrt{R}}\right)$ we get

$$\langle Tx|x\rangle_H = \langle\sqrt{\sqrt{R}S^{-1}}\sqrt{\sqrt{R}}x|x\rangle_H = \langle S^{-1}\sqrt{\sqrt{R}}x|\sqrt{\sqrt{R}}x\rangle_H \geq 0,$$

due to the accretivity of S^{-1}. Since $T = \sqrt{\sqrt{R}S^{-1}}\sqrt{\sqrt{R}}$, we deduce the accretivity of T. To show that T is a root, it suffices to check $T^2 = A$. Using $\sqrt{R}S^{-1} \subseteq S^{-1}\sqrt{R}$, we get

$$T^2 = S^{-1}\sqrt{R}S^{-1}\sqrt{R}$$
$$\subseteq S^{-1}S^{-1}\sqrt{R}\sqrt{R}$$
$$= \left(S^2\right)^{-1}R$$
$$= (1+A)\left(1 - (1+A)^{-1}\right)$$
$$= 1 + A - 1$$
$$= A.$$

However, since $T^2 = T^*T$ and A are both selfadjoint and accretive, we deduce that

$$\left(1 + T^2\right)^{-1} \subseteq (1+A)^{-1}$$

which in turn yields

$$\left(1 + T^2\right)^{-1} = (1+A)^{-1},$$

since $(1 + T^2)^{-1}$ is already defined on the whole Hilbert space H. The latter implies $T^2 = A$, that is, T is a root of A.

We now show the uniqueness of the root of A. Assume there is another selfadjoint accretive operator $C : \text{dom}(C) \subseteq H \to H$ with $C^2 = A$. Then clearly $CA = AC$, which in turn implies

$$(1 + C)^{-1}(1 + A)^{-1} = (1 + A)^{-1}(1 + C)^{-1}.$$

Thus, by Theorem B.8.2,

$$(1 + C)^{-1} S = S(1 + C)^{-1}$$

and hence,

$$CS = \overline{SC} = \overline{\sqrt{S} C \sqrt{S}}$$

according to Lemma B.8.3 (with $\lambda = 1$ and $B = S$). As above, this implies that CS is a selfadjoint accretive operator. Moreover,

$$(CS)(CS) \subseteq C^2 S^2 = A(1 + A)^{-1} = R.$$

In particular

$$CS(1 + CS) \subseteq CS + R$$

and hence,

$$CS = CS(1 + CS)(1 + CS)^{-1} \subseteq CS(1 + CS)^{-1} + R(1 + CS)^{-1} \in \mathcal{B}(H),$$

which implies, since CS is densely defined and closed, that $CS \in \mathcal{B}(H)$. Since $(CS)(CS) = R$ and CS is selfadjoint and accretive, we deduce that $CS = \sqrt{R}$ by Theorem B.8.2. Thus,

$$(1 + T)^{-1} = \left(1 + S^{-1}\sqrt{R}\right)^{-1} = \left(1 + S^{-1}CS\right)^{-1} = S^{-1}(1 + C)^{-1} S = (1 + C)^{-1},$$

which shows $T = C$. Hence, the root of A is uniquely determined.   □

With the root of an accretive selfadjoint operator at hand, we can come to the polar decomposition of a densely defined closed linear operator. We need the following auxiliary result.

Lemma B.8.5 *Let $A : \text{dom}(A) \subseteq H_0 \to H_1$ be a densely defined closed linear operator. Then $\overline{A|_{\text{dom}(A^*A)}} = A$.*

Proof Since A is closed, its domain $\mathrm{dom}(A)$ is a Hilbert space with respect to the graph inner product

$$\langle u|v\rangle_{\mathrm{dom}(A)} = \langle u|v\rangle_{H_0} + \langle Au|Av\rangle_{H_1} \quad (u, v \in \mathrm{dom}(A)).$$

We note that the assertion is equivalent to the density of $\mathrm{dom}(A^*A)$ in $\mathrm{dom}(A)$ with respect to the graph inner product. The latter is equivalent to

$$\mathrm{dom}(A^*A)^{\perp \mathrm{dom}(A)} = \{0\}$$

according to Corollary B.2.6. Let $x \in \mathrm{dom}(A^*A)^{\perp \mathrm{dom}(A)}$, that is, $x \in \mathrm{dom}(A)$ and for all $y \in \mathrm{dom}(A^*A)$

$$\langle x|y\rangle_{\mathrm{dom}(A)} = 0.$$

Then

$$0 = \langle x|y\rangle_{H_0} + \langle Ax|Ay\rangle_{H_1} = \langle x|y\rangle_{H_0} + \langle x|A^*Ay\rangle_{H_0} = \langle x|(1 + A^*A)y\rangle_{H_0}$$

for all $y \in \mathrm{dom}(A^*A)$. Hence, $x \in ((1 + A^*A)[H])^{\perp} = [\{0\}](1+A^*A)$ by Theorem B.4.8 and Proposition B.4.17. Thus,

$$|x|^2_{H_0} \leq \langle x|x\rangle_{H_0} + \langle Ax|Ax\rangle_{H_1} = \langle x|(1 + A^*A)x\rangle_{H_0} = 0,$$

which implies $x = 0$. Thus, $\mathrm{dom}(A^*A)$ is dense in $\mathrm{dom}(A)$ with respect to the graph inner product.   □

Proposition B.8.6 (Polar Decomposition) *Let* $A : \mathrm{dom}(A) \subseteq H_0 \to H_1$ *be a densely defined closed linear operator. We set*

$$|A| := \sqrt{A^*A},$$

which exists according to Proposition B.4.17 and Theorem B.8.4. Then there exists a unitary operator $U : \overline{|A|[H_0]} \to \overline{A[H_0]}$ *such that*

$$A = \iota_\mathrm{r}(A)U\iota_\mathrm{r}(|A|)^*|A|,$$

where $\iota_\mathrm{r}(A) := \iota_{\overline{A[H_0]}} : \overline{A[H_0]} \hookrightarrow H_1$ *denotes the canonical embedding (similarly for* $\iota_\mathrm{r}(|A|)$*).*

Proof Let $x \in \text{dom}(A^*A) = \text{dom}(|A|^2)$. Then we compute

$$|Ax|_{H_1}^2 = \langle Ax | Ax \rangle_{H_1} = \langle x | A^*Ax \rangle_{H_0} = \langle x | |A|^2 x \rangle_{H_0} = \langle |A|x | |A|x \rangle_{H_0} = |||A|x|_{H_0}^2,$$

which implies

$$|Ax|_{H_1} = |||A|x|_{H_0} \quad (x \in \text{dom}(A^*A)). \tag{B.8.2}$$

Let $x \in \text{dom}(A)$. Since $A = \overline{A|_{\text{dom}(A^*A)}}$ by Lemma B.8.5, there exists a sequence $(x_n)_{n \in \mathbb{N}}$ in $\text{dom}(A^*A)$ with $x_n \to x$ and $Ax_n \to Ax$. From (B.8.2), we get that $(|A|x_n)_{n \in \mathbb{N}}$ is a Cauchy-sequence and thus, convergent. Due to the closedness of $|A|$, we derive $x \in \text{dom}(|A|)$ and $|A|x = \lim_{n \to \infty} |A|x_n$. Thus $\text{dom}(A) \subseteq \text{dom}(|A|)$ and $|Ax|_{H_1} = |||A|x|_{H_0}$ for all $x \in \text{dom}(A)$. Since in addition $|A| = \overline{|A||_{\text{dom}(|A|^2)}} = \overline{|A||_{\text{dom}(A^*A)}}$ by Lemma B.8.5, the same rationale yields $\text{dom}(A) = \text{dom}(|A|)$ and

$$|Ax|_{H_1} = |||A|x|_{H_0} \quad (x \in \text{dom}(A) = \text{dom}(|A|)). \tag{B.8.3}$$

Now, define

$$U : |A|[H_0] \subseteq \overline{|A|[H_0]} \to \overline{A[H_0]}$$

$$|A|x \mapsto Ax.$$

Then U is a well-defined isometry by (B.8.3). Moreover, U has dense range and thus, it can be extended to a unitary operator $U : \overline{|A|[H_0]} \to \overline{A[H_0]}$. For $x \in \text{dom}(|A|) = \text{dom}(A)$ we get

$$\iota_r(A) U \iota_r(|A|)^* |A|x = Ax,$$

which shows the last part of the assertion. $\square$

Proposition B.8.7 *Let* $A : \text{dom}(A) \subseteq H_0 \to H_1$ *be a densely defined closed linear operator. Moreover, let* $U : \overline{|A|[H_0]} \to \overline{A[H_0]}$ *be unitary with*

$$A = \widetilde{U} |A|,$$

where $\widetilde{U} := \iota_r(A) U \iota_r(|A|)^* \in \mathcal{B}(H_0, H_1)$. *Then*

$$\sqrt{AA^*} = |A^*| = \widetilde{U} |A| \widetilde{U}^*$$

Proof First, note that

$$A^* = (\widetilde{U}|A|)^* = |A|\widetilde{U}^* \tag{B.8.4}$$

by Proposition B.4.10. We claim that $\widetilde{U}|A|\widetilde{U}^*$ is selfadjoint and accretive. Indeed, accretivity immediately follows from the accretivity of $|A|$ and

$$\left(\widetilde{U}|A|\widetilde{U}^*\right)^* = \left(\widetilde{U}A^*\right)^* = A\widetilde{U}^* = \widetilde{U}|A|\widetilde{U}^*,$$

that is, $\widetilde{U}|A|\widetilde{U}^*$ is selfadjoint. Moreover, using that $\widetilde{U}^*\widetilde{U} = P_{\overline{|A|[H_0]}}$, we obtain

$$AA^* = \widetilde{U}|A||A|\widetilde{U}^*$$
$$= \left(\widetilde{U}|A|\widetilde{U}^*\right)\left(\widetilde{U}|A|\widetilde{U}^*\right).$$

Hence, by the uniqueness of the root, the assertion follows.   □

Corollary B.8.8 *Let* $A : \mathrm{dom}(A) \subseteq H_0 \to H_1$ *be a densely defined closed linear operator. Then*

$$A|A| = |A^*|A.$$

Proof Since $A = \widetilde{U}|A|$, we obtain $A^* = |A|\widetilde{U}^*$ and hence,

$$A|A| = \widetilde{U}|A|^2 = \widetilde{U}A^*A = \widetilde{U}|A|\widetilde{U}^*A = |A^*|A,$$

where we have used Proposition B.8.7.   □

Bibliography

1. W. Arendt, C.J.K. Batty, M. Hieber, F. Neubrander, *Vector-Valued Laplace Transforms and Cauchy Problems*. Monographs in Mathematics, vol. 96 (Birkhäuser, Basel, 2001), xi, 523 p.
2. G. Barbatis, I.G. Stratis, Homogenization of Maxwell's equations in dissipative bianisotropic media. Math. Method. Appl. Sci. **26**, 1241–1253 (2003)
3. C. Batty, L. Paunonen, D. Seifert, Optimal energy decay in a one-dimensional coupled wave-heat system. J. Evol. Eq. **16**(3), 649–664 (2016)
4. A. Bertram, *Elasticity and Plasticity of Large Deformations: An Introduction* (Springer, Berlin, 2005), xiv, 326 p
5. M.A. Biot, General theory of three-dimensional consolidation. J. Appl. Phys. Lancaster Pa. **12**, 155–164 (1941)
6. M.A. Biot, Thermoelasticity and irreversible thermodynamics. J. Appl. Phys. **27**, 240–253 (1956)
7. H. Brézis, A. Haraux, Image d'une somme d'opérateurs monotones et applications. Isr. J. Math. **23**, 165–186 (1976)
8. R. Chill, E. Fasangova, *Gradient Systems*. Lecture Notes of the 13th International Internet Seminar (Matfyzpress, Prague, 2010)
9. A.J. Chorin, A numerical method for solving incompressible viscous flow problems [J. Comput. Phys. **2** (1967), no. 1, 12–36]. J. Comput. Phys. **135**(2), 115–125 (1997). With an introduction by Gerry Puckett, Commemoration of the 30th anniversary {of J. Comput. Phys.}
10. G. Da Prato, P. Grisvard, Sommes d'operateurs lineaires et equations differentielles operationnelles. J. Math. Pures et Appl. **54**, 305–387 (1975)
11. N. Dunford, J.T. Schwartz, *Linear Operators*. Spectral Theory, vol. II (Wiley, Hoboken, 1963)
12. K. Engel, R. Nagel, *One-Parameter Semigroups for Evolution Equations*, vol. 194 (Springer, New York, 1999)
13. S. Franz, M. Waurick, Resolvent estimates and numerical implementation for the homogenisation of one-dimensional periodic mixed type problems. Z. Angew. Math. Mech. **98**(7), 1284–1294 (2018)
14. S. Franz, S. Trostorff, M. Waurick, Numerical methods for changing type systems. Technical report. IMA J. Numer. Anal. **39**(2), 1009–1038 (2019)
15. K.O. Friedrichs, On the boundary value problems of the theory of elasticity and korn's inequality. Ann. Math. **48**, 441–471 (1947)
16. K.O. Friedrichs, Symmetric hyperbolic linear differential equations. Comm. Pure Appl. Math. **7**, 345–392 (1954)

© Springer Nature Switzerland AG 2020

R. Picard et al., *A Primer for a Secret Shortcut to PDEs of Mathematical Physics*, Frontiers in Mathematics, https://doi.org/10.1007/978-3-030-47333-4

17. G.P. Galdi, *An Introduction to the Mathematical Theory of the Navier-Stokes Equations. Vol. 1: Linearized Steady Problems*. Springer Tracts in Natural Philosophy, vol. 38 (Springer, New York, 1994)

18. R.A. Guyer, J.A. Krumhansl, Dispersion relation for second sound in solids. Phys. Rev. **133**, A1411–A1417 (1964)

19. R.A. Guyer, J.A. Krumhansl, Solution of the linearized phonon boltzmann equation. Phys. Rev. **148**, 766–778 (1966)

20. R.A. Guyer, J.A. Krumhansl, Thermal conductivity, second sound, and phonon hydrodynamic phenomena in nonmetallic crystals. Phys. Rev. **148**, 778–788 (1966)

21. E. Hille, R.S. Phillips, *Functional Analysis and Semigroups*. American Mathematical Society Colloquium Publications, vol. 31 (American Mathematical Society, Providence, 1957)

22. A. Kalauch, R. Picard, S. Siegmund, S. Trostorff, M. Waurick, A hilbert space perspective on ordinary differential equations with memory term. J. Dyn. Differ. Eq. **26**(2), 369–399 (2014)

23. T. Kato, *Perturbation Theory for Linear Operators, Corrected Printing of the 2nd ed.* Grundlehren der mathematischen Wissenschaften, vol. 132, XXI (Springer, Berlin, 1980)

24. A. Korn, Die Eigenschwingungen eines elastischen Körpers mit ruhender Oberfläche. Akad. Wiss. München, Math. Phys. Kl. Ber. **36**, 351–401 (1906)

25. A. Korn, Über einige Ungleichungen, welche in der Theorie der elastischen und elektrischen Schwingungen eine Rolle spielen. Bull. Int. Cracovie Akad. Umiejet, Classe Sci. Math. Nat. **3**, 705–724 (1909)

26. R. Leis, *Initial Boundary Value Problems in Mathematical Physics* (Wiley and B.G. Teubner, Stuttgart, 1986)

27. H.W. Lord, Y. Shulman, A generalized dynamical theory of thermoelasticity. J. Mech. Phys. Solids **15**(5), 299–309 (1967)

28. J.F. Lu, A. Hanyga, D.S. Jeng, A linear dynamic model for a saturated porous medium. Trans. Porous Media **68**, 321–340 (2007)

29. J.C. Maxwell, *A Treatise on Electricity and Magnetism. Vol. 1*. Oxford Classic Texts in the Physical Sciences (The Clarendon Press, Oxford University Press, Oxford, 1998). With prefaces by W. D. Niven and J. J. Thomson, Reprint of the third (1891) edition

30. J.C. Maxwell, *A Treatise on Electricity and Magnetism. Vol. 2*. Oxford Classic Texts in the Physical Sciences (The Clarendon Press, Oxford University Press, New York, 1998). Reprint of the third (1891) edition

31. D.F. McGhee, R. Picard, A note on anisotropic, inhomogeneous, poro-elastic media. Math. Method. Appl. Sci. **33**(3), 313–322 (2010)

32. D. McGhee, R. Picard, S. Trostorff, M. Waurick, *Mathematical Transformations*, ch. 15 (Wiley, Hoboken, 2014), pp. 503–526

33. R.D. Mindlin. Equations of high frequency vibrations of thermopiezoelectric crystal plates. Int. J. Solids Struct. **10**, 625–637 (1974)

34. C.S. Morawetz, The limiting amplitude principle. Comm. Pure Appl. Math. 15, 349–361 (1962)

35. C.S. Morawetz, The limiting amplitude principle for arbitrary finite bodies. Comm. Pure Appl. Math. **18**, 183–189 (1965)

36. D. Morgenstern, Beiträge zur nichtlinearen Funktionalanalysis. PhD Thesis, TU Berlin, 1952

37. S. Mukhopadyay, R. Picard, S. Trostorff, M. Waurick, On some models in linear thermoelasticity with rational material laws. Math. Mech. Solids **21**(9), 1149–1163 (2016)

38. S. Mukhopadyay, R. Picard, S. Trostorff, M. Waurick, A note on a two-temperature model in linear thermoelasticity. Math. Mech. Solids **22**(5), 905–918 (2017)

39. A.J. Mulholland, R. Picard, S. Trostorff, M. Waurick. On well-posedness for some thermo-piezo-electric coupling models. Math. Meth. Appl. Sci. **39**(15), 4375–4384 (2016)

40. NASA, Negative dielectric constant material. https://technology.nasa.gov/patent/LAR-TOPS-177
41. W. Nowacki, Dynamische probleme der unsymmetrischen Elastizität. Prikl. Mekh. **6**(4), 31–50 (1970)
42. W. Nowacki, *Theory of Asymmetric Elasticity* (Pergamon Press/PWN-Polish Scientific Publishers, Oxford/Warszawa, 1986), VIII, 383 p
43. J.W. Nunziato, S.C. Cowin, A nonlinear theory of elastic materials with voids. Arch. Ration. Mech. Anal. **72**, 175–201 (1979)
44. D. Pauly, Low frequency asymptotics for time-harmonic generalized Maxwell's equations in nonsmooth exterior domains. Adv. Math. Sci. Appl. **16**(2), 591–622 (2006)
45. D. Pauly, Complete low frequency asymptotics for time-harmonic generalized Maxwell equations in nonsmooth exterior domains. Asymptotic Anal. **60**(3–4), 125–184 (2008)
46. D. Pauly, W. Zulehner, On closed and exact grad grad- and div div-complexes, corresponding compact embeddings for tensor rotations, and a related decomposition result for Biharmonic problems in 3D. Technical Report, University Duisburg–Essen, 2017. https://arxiv.org/abs/1609.05873v2
47. I. Petrowsky, On the diffusion of waves and the lacunas for hyperbolic equations. Rec. Math. [Mat. Sbornik] N. S. **17**(59), 289–370 (1945)
48. R. Picard, Ein Hodge-Satz für Mannigfaltigkeiten mit nicht-glattem Rand. Math. Methods Appl. Sci. **5**, 153–161 (1983)
49. R. Picard, On the low frequency asymptotics in electromagnetic theory. J. Reine Angew. Math. **354**, 50–73 (1984)
50. R. Picard, The low frequency limit for time-harmonic acoustic waves. Math. Methods Appl. Sci. **8**, 436–450 (1986)
51. R. Picard, *Hilbert Space Approach to Some Classical Transforms* (Wiley, New York, 1989)
52. R. Picard, Some decomposition theorems and their application to nonlinear potential theory and Hodge theory. Math. Methods Appl. Sci. **12**(1), 35–52 (1989)
53. R. Picard, On a selfadjoint realization of curl and some of its applications. Ric. Mat. **47**(1), 153–180 (1998)
54. R. Picard, On a selfadjoint realization of curl in exterior domains. Math. Z. **229**(2), 319–338 (1998)
55. R. Picard, Evolution equations as operator equations in lattices of Hilbert spaces. Glas. Mat. Ser. III **35**, 111–136 (2000)
56. R. Picard, The Stokes system in the incompressible case—revisited, in ed. by Rencławowicz, J., et al., *Parabolic and Navier-Stokes Equations. Part 2. Proceedings of the confererence, Będlewo, Poland, September 10–17, 2006, Banach Center Publications.*, vol. 81 (Polish Academy of Sciences, Institute of Mathematics, Warsaw, 2008), pp. 369–382
57. R. Picard, A structural observation for linear material laws in classical mathematical physics. Math. Methods Appl. Sci. **32**(14), 1768–1803 (2009)
58. R. Picard, A structural observation for linear material laws in classical mathematical physics. Math. Methods Appl. Sci. **32**, 1768–1803 (2009)
59. R. Picard, A class of evolutionary problems with an application to acoustic waves with impedance type boundary conditions, in ed. by W. Arendt et al. *Spectral Theory, Mathematical System Theory, Evolution Equations, Differential and Difference Equations. Selected Papers of 21st International Workshop on Operator Theory and Applications, IWOTA10, Berlin, Germany, July 12–16, 2010.* Operator Theory: Advances and Applications, vol. 221 (Birkhäuser, Basel, 2012), pp. 533–548.
60. R. Picard, Mother operators and their descendants. J. Math. Anal. Appl. **403**(1), 54–62 (2013)

61. R. Picard, On well-posedness for a Piezo-electromagnetic coupling model with boundary dynamics. Comput. Methods Appl. Math. **17**(3), 499–513 (2017)
62. R. Picard, D.F. McGhee, *Partial Differential Equations: A Unified Hilbert Space Approach*. De Gruyter Expositions in Mathematics, vol. 55 (De Gruyter, Berlin, 2011), 518 p
63. R. Picard, S. Trostorff, M. Waurick, M. Wehowski, On non-autonomous evolutionary problems. J. Evol. Eq. **13**(4), 751–776 (2013)
64. R. Picard, S. Trostorff, M. Waurick, A functional analytic perspective to delay differential equations. Oper. Matrices **8**(1), 217–236 (2014)
65. R. Picard, S. Trostorff, M. Waurick, On a connection between the Maxwell system, the extended Maxwell system, the Dirac operator and gravito-electromagnetism. Math. Meth. Appl. Sci. 40, 415–434 (2014)
66. R. Picard, S. Trostorff, M. Waurick, On evolutionary equations with material laws containing fractional integrals. Math. Meth. Appl. Sci. **38**(15), 3141–3154 (2015)
67. R. Picard, S. Trostorff, M. Waurick, On some models for elastic solids with micro-structure. Zeitschrift für Angew. Math. Mech. **95**(7), 664–689 (2015)
68. R. Picard, S. Trostorff, M. Waurick, Well-posedness via monotonicity. an overview, in *Operator Semigroups Meet Complex Analysis, Harmonic Analysis and Mathematical Physics. Operator Theory: Advances and Applications*, vol. 250 (Birkhäuser, Cham, 2015), pp. 397–452
69. R. Picard, S. Seidler, S. Trostorff, M. Waurick, On abstract grad-div systems. J. Differ. Eq. **260**(6), 4888–4917 (2016)
70. R. Picard, S. Trostorff, M. Waurick, On a comprehensive class of linear control problems. IMA J. Math. Control Inf. **33**(2), 257–291 (2016)
71. R. Picard, S. Trostorff, M. Waurick, On Maximal regularity for a class of evolutionary equations. J. Math. Anal. Appl. **449**(2), 1368–1381 (2017)
72. R. Picard, S. Trostorff, M. Waurick, On the well-posedness of a class of non-autonomous SPDEs: an operator-theoretical perspective . GAMM-Mitteilungen **41**(4), e201800014 (2018). Applied Operator Theory - Part II
73. M. Reed, B. Simon, *Methods of Modern Mathematical Physics. I: Functional Analysis (Rev. and enl. ed.)* (Academic/A Subsidiary of Harcourt Brace Jovanovich, New York/San Diego, 1980)
74. R. Seeliger, *Elektronentheorie der Metalle*. Encyklopädie der mathematischen Wissenschaften. Mit Einschluß ihrer Anwendungen., vol. 5 (Physik) (Teubner, Leipzig, 1922), pp. 777–878
75. C. Seifert, S. Trostorff, M. Waurick, *Evolutionary Equations*. Lecture Notes for the 23rd Internet Seminar (2020). http://www.mat.tuhh.de/isem23
76. A. Süß, M. Waurick, A solution theory for a general class of SPDEs. Stochastics Partial Differ. Eq.: Anal. Comput. **5**(2), 278–318 (2017)
77. M. Taskinen, S. Vänskä, Current and charge integral equation formulations and picard's extended maxwell system. IEEE Trans. Antenn. Propag. **55**, 3495–3503 (2007)
78. A.F.M. ter Elst, G. Gorden, M. Waurick, The Dirichlet-to-Neumann operator for divergence form problems. Annali di Matematica Pura ed Applicata, **198**(1): 177–203, (2019)
79. S. Trostorff, An alternative approach to well-posedness of a class of differential inclusions in Hilbert spaces. Nonlinear Anal. Theory Methods Appl. Ser. A Theory Methods **75**(15), 5851–5865 (2012)
80. S. Trostorff, A characterization of boundary conditions yielding maximal monotone operators. J. Funct. Anal. **267**(8), 2787–2822 (2014)
81. S. Trostorff, Exponential stability and initial value problems for evolutionary equations. Habilitation, TU Dresden, 2018
82. S. Trostorff, M. Waurick, A note on elliptic type boundary value problems with maximal monotone relations. Math. Nachr. **287**(13), 1545–1558 (2014)

83. S. Trostorff, M. Wehowski, Well-posedness of non-autonomous evolutionary inclusions. Non-linear Anal. **101**, 47–65 (2014)

84. P. Vernotte, Les paradoxes de la theorie continue de l'equation de la chaleur. C.R. Acad. Sci. Paris II **246**, 3154 (1958)

85. M. Waurick, Limiting processes in evolutionary equations - A Hilbert space approach to homogenization. Dissertation, Technische Universität Dresden, 2011. http://nbn-resolving.de/urn:nbn:de:bsz:14-qucosa-67442.

86. M. Waurick, A note on causality in Banach spaces. Indagat. Math. **26**(2), 404–412 (2015)

87. M. Waurick, On non-autonomous integro-differential-algebraic evolutionary problems. Math. Methods Appl. Sci. **38**(4), 665–676 (2015)

88. M. Waurick, On the continuous dependence on the coefficients of evolutionary equations. Habilitation, Technische Universität Dresden, 2016. http://arxiv.org/abs/1606.07731.

89. M. Waurick, On the homogenization of partial integro-differential-algebraic equations. Oper. Matrices **10**(2), 247–283 (2016)

90. M. Waurick, Stabilization via homogenization. Appl. Math. Lett. **60**, 101–107 (2016)

91. M. Waurick, Nonlocal H-convergence. Calc. Var. Partial Differ. Eq. **57**, 46 (2018)

92. N. Wellander, Homogenization of the Maxwell equations: case I. linear theory. Appl. Math. **46**, 29–51 (2001)

Index

© Springer Nature Switzerland AG 2020

R. Picard et al., *A Primer for a Secret Shortcut to PDEs of Mathematical Physics*,
Frontiers in Mathematics, https://doi.org/10.1007/978-3-030-47333-4

Printed in the United States
By Bookmasters